Lecture Notes in Mathematics 2012

Editors:
J.-M. Morel, Cachan
B. Teissier, Paris

Subseries:
École d'Été de Probabilités de Saint-Flour

Saint-Flour Probability Summer School

The Saint-Flour volumes are reflections of the courses given at the Saint-Flour Probability Summer School. Founded in 1971, this school is organised every year by the Laboratoire de Mathématiques (CNRS and Université Blaise Pascal, Clermont-Ferrand, France). It is intended for PhD students, teachers and researchers who are interested in probability theory, statistics, and in their applications.

The duration of each school is 13 days (it was 17 days up to 2005), and up to 70 participants can attend it. The aim is to provide, in three high-level courses, a comprehensive study of some fields in probability theory or Statistics. The lecturers are chosen by an international scientific board. The participants themselves also have the opportunity to give short lectures about their research work.

Participants are lodged and work in the same building, a former seminary built in the 18th century in the city of Saint-Flour, at an altitude of 900 m. The pleasant surroundings facilitate scientific discussion and exchange.

The Saint-Flour Probability Summer School is supported by:

– Université Blaise Pascal
– Centre National de la Recherche Scientifique (C.N.R.S.)
– Ministère délégué à l'Enseignement supérieur et à la Recherche

For more information, see back pages of the book and
http://math.univ-bpclermont.fr/stflour/

Jean Picard
Summer School Chairman
Laboratoire de Mathématiques
Université Blaise Pascal
63177 Aubière Cedex
France

Alison Etheridge

Some Mathematical Models from Population Genetics

École d'Été de Probabilités
de Saint-Flour XXXIX-2009

 Springer

Alison Etheridge
University of Oxford
Department of Statistics
1 South Parks Road
Oxford OX1 3TG
United Kingdom
etheridg@stats.ox.ac.uk

ISBN: 978-3-642-16631-0   e-ISBN: 978-3-642-16632-7
DOI: 10.1007/978-3-642-16632-7
Springer Heidelberg Dordrecht London New York

Lecture Notes in Mathematics ISSN print edition: 0075-8434
ISSN electronic edition: 1617-9692

Cover design: SPi Publisher Services

Printed on acid-free paper

Springer is part of Springer Science+Business Media (www.springer.com)

Preface

These notes on mathematical models from population genetics reflect the 16 h of lectures that I delivered in St Flour in July 2009. Other than minor corrections and clarifications, they have changed very little in the year that has elapsed since then. Although it was tempting to add more material, I concluded that not only would this lead to unacceptable delays, but it would also be redundant. Whereas there are few references of which I am aware that present the material covered here in a self-contained way, there are now many texts that cover, for example, coalescent theory in more detail.

The notes are intended for graduate students in mathematics. They aim to introduce the reader to a range of mathematical models that have their origins in theoretical population genetics. Some date right back to the origins of the subject and some were introduced in the last few years. All share a rich mathematical structure. Research on the more recent models, notably the Λ-coalescents and their spatial analogues, is progressing at a breathtaking speed and so it is impossible to provide a comprehensive survey of what is known. Instead I have aimed to explain some of the reasons that such models are interesting biologically and to equip the reader with enough background to be able to browse the literature as it appears.

There are many people to whom I owe thanks. My interest in population genetics stems from a collaboration with Nick Barton (IST Austria and the University of Edinburgh). Working with Nick over many years has been a privilege and a pleasure and almost all the material covered here I first learned about through conversations with him. I am grateful to Leif Döring, Bjarki Eldon, Bob Griffiths, Habib Saadi and the many others who read and commented on parts of the manuscript. Special thanks are due to Amandine Véber, who went through several iterations of the whole document in tremendous detail and undoubtedly improved the notes beyond recognition. I was fortunate to spend the first four months of 2009 visiting Université Paris Sud in Orsay. My thanks go to everyone there, especially Yves Le Jan for making that possible. While I was in Orsay, Jean-François Le Gall persuaded me to give a masters course as a dry run for (at least part of) this course. The experience was extremely valuable and my thanks go to Jean-François and to the enthusiastic audience. Jean Picard quietly ensured that everything at St Flour ran extremely smoothly and the participants were tremendous. I simply had a lot of fun. Finally, as always, I thank Lionel, Charlotte and Matthew Mason for all their support and understanding.

Oxford, August 2010 *Alison Etheridge*

Contents

Chapter 1
Introduction

The main purpose of theoretical population genetics is to understand the complex patterns of genetic variation that we observe in the world around us. Its origins can be traced to the pioneering work of Fisher, Haldane and Wright. Their contributions were fundamental in establishing the *Modern Evolutionary Synthesis*, in which Darwin's theory of evolution by natural selection was finally reconciled with Mendelian genetics. Darwin's theory of evolution Darwin (1859) can be simply stated: "Heritable traits that increase reproductive success will become more common in a population". Thus, in order for natural selection to act, there must be variation within a population and offspring must be similar to their parents. So to fully understand evolution we need a mechanism whereby variation is created and inherited. This is provided by Mendelian genetics Mendel (1866). Again the idea can be simply stated. Traits are determined by *genes*. Each gene occurs in finitely many different types that we call *alleles* and different alleles may produce different traits. Offspring are similar to their parents because they inherit genes from their parents. The difficulty is that Darwin had argued that evolution of complex, well-adapted organisms depends on selection acting on a large number of slight variants in a trait and much of Mendel's work deliberately focused on discontinuous changes in traits determined by a single gene.

The resolution lay in the foundations of theoretical population genetics. In 1918, Fisher showed how correlations between relatives that had been measured by biometricians could be explained by multiple Mendelian factors together with random, non-genetic, influences. In the process he developed the statistical theory of *analysis of variance*. He went on to show that Mendelian genetics was consistent with the idea of evolution by natural selection. Thus, if traits depend on multiple genes, each making a small contribution, the apparently discontinuous nature of Mendelian inheritance is reconciled with continuous variation and gradual evolution. Starting in 1924, Haldane published a series of papers that provide a detailed theoretical analysis of how differences in survival or reproduction due to one or two Mendelian genes would affect a population. He used examples like the evolution of the peppered

A. Etheridge, *Some Mathematical Models from Population Genetics*, Lecture Notes in Mathematics 2012, DOI 10.1007/978-3-642-16632-7_1,

moth to show that natural selection could act extremely fast.[1] In a series of papers starting in 1922, Wright quantified the way in which the random process of reproduction in a finite population would lead to changes in allele frequency and examined how this *random genetic drift* interacted with selection, mutation and migration. He introduced the notion of an adaptive landscape in which natural selection would drive a population towards a local maximum, but genetic drift could push the population away from such a peak paving the way for natural selection to push it towards a different peak. Through this mechanism the population explores the evolutionary landscape. Mathematical modelling played a crucial rôle in the work of all three men.

Of course many details remained (and indeed remain) unclear. For example, what is the relative importance of mutation, selection, random drift and population subdivision for the genetic variation observed today? Fisher emphasised gradual changes in a single large population due to selection acting on small variants, Haldane placed more importance on strong selection acting on single genes and Wright argued that adaptation would be most effective in a population that was subdivided into many small subpopulations in his *shifting balance* theory.

At the time of the evolutionary synthesis, genetic variability could not be observed directly. Early work was restricted to genes that happened to be detectable in an observable way, but over the subsequent fifty years things changed dramatically. In 1953, armed with key physical evidence obtained by Rosalind Franklin and Maurice Wilkins, Watson and Crick showed that DNA forms a double helix in which one sequence of bases pairs with the complementary sequence.[2] Over the next decade it was established how DNA codes for proteins through the genetic code. Beginning in the mid-1960s scientists began to study genetic variation at the scale of DNA, RNA[3] and proteins. Studies revealed an unexpectedly high level of variation within species. It was also possible to compare the evolution of the same protein across different species. By comparing species that diverged from one another at a known time, it was found that any given protein had evolved at a steady rate, even if it was evolving in very different organisms. In other words, there is a *molecular clock*. In 1968, Kimura famously (and controversially) argued that there

---

[1] The peppered moth was originally predominantly light-coloured, providing camouflage on the lichen coloured trees on which it rests. As a result of pollution the lichens died out and the trees became blackened by soot, making the light coloured moths vulnerable to predation, and dark-coloured moths flourished. With improving environmental conditions, light coloured moths have once again become common.

[2] DNA stands for deoxyribonucleic acid. The four bases are A (adenine), T (thymine), C (cytosine) and G (guanine) and A bonds to T and C to G.

[3] RNA is similar to DNA. It consists of a chain of nucleotides. Messenger RNA (mRNA) carries information from DNA to the ribosome where proteins are synthesised. The coding sequence of the mRNA determines the amino acid sequence in the protein that is produced. Every three nucleotides (a codon) code for one amino acid. This *genetic code* has redundancy, but no ambiguity. The $4^3 = 64$ different triplets code for 20 different amino acids and a stop codon.

Like DNA, RNA can carry genetic information. For example RNA viruses have genomes composed of RNA.

is too much genetic variation in species for more than a small fraction to be subject to natural selection. He also argued that the molecular clock was best explained by the steady accumulation of mutations that have no effect on fitness. Others, by contrast, were emphasising the rôle of selection in explaining even very small changes in phenotype. The theory remains controversial. It is simply not known what proportion of differences between individuals or species are maintained by selection, but Kimura's theory provides a valuable 'null' model against which data can be tested.

If we are to assess the relative importance of mutation, selection, drift, spatial structure and so on, then the first step is to distill our understanding of how these processes operate into a workable mathematical model whose predictions can be compared to data. Even now we don't know exactly how genes combine to influence a whole organism or what maintains the variation in those genes. But with advances in molecular biology, a wealth of data is available and mathematicians have a key rôle to play. Over the last three decades, in parallel with advances in DNA sequencing technology, new mathematical models have been introduced focusing on the genealogy (that is the ancestral history) of a random sample of genes from a population. These 'coalescent' models have a rich and beautiful mathematical structure and in addition to providing the necessary tools for the interpretation of genetic data they have become a popular playground for mathematicians.

In these lectures we will introduce and study some models (both old and new) that have their origins in theoretical population genetics. We will try to minimise the use of biological jargon, but we end this section with a note on terminology. The 'atom' of genetics is the single base pair or *nucleotide*. It is often referred to as a *site*. The term *locus* is used to refer in a general way to a location in the genome. It may refer to anything from a few hundred bases to a long stretch of DNA containing several genes. Whereas in classical Mendelian genetics a gene was a single well-defined unit, it is now loosely defined as a stretch of DNA that includes sequences that code for a protein (or a functional RNA molecule) and regulatory sequences. The genes are organised on chromosomes and, for mathematical convenience, we shall consider chromosomes to be linear. An excellent introduction to the underlying biology is Barton et al. (2007).

Chapter 2
Mutation and Random Genetic Drift

2.1 The Wright–Fisher Model and the Kingman Coalescent

Evolution is a random process. Random events enter in many ways, from errors in copying genetic material to small and large scale environmental changes, but the most basic source of randomness that we must understand is due to reproduction in a finite population leading to *random genetic drift*. The simplest model of random genetic drift was developed independently by Sewall Wright and R.A. Fisher and is known as the Wright–Fisher model. We consider a population in which every individual is equally likely to mate with every other and in which all individuals experience the same conditions. Such a population is called *panmictic*. We also suppose that the population is *neutral* (everyone has an equal chance of reproductive success). Most species are either *haploid* meaning that they have a single copy of each chromosome (for example, most bacteria), or *diploid* meaning that they have two copies of each chromosome (for example, humans). We suppose that the population is haploid, so that each individual has exactly one parent. Although in a diploid population individuals have two parents, each *gene* can be traced to a single parental gene in the previous generation and so it is customary in this setting to model the genes in a diploid population of size N as a haploid population of size $2N$.[1] As we shall see in Sect. 5.6, this device fails once we are interested in tracing several genes at the same time.

Definition 2.1 (The neutral Wright–Fisher model). The neutral Wright–Fisher model for a panmictic, haploid population of constant size N is described as follows. The population of N genes evolves in discrete generations. Generation $(t+1)$ is formed from generation t by choosing N genes uniformly at random with replacement. That is, each gene in generation $(t+1)$ chooses its parent independently at random from those present in generation t.

---

[1] In fact we are assuming that the population is hermaphrodite here – so there are no separate sexes – and we are allowing a small chance of self-fertilisation. For a population subdivided into N_m males and N_f females we can still use the same model, but with an *effective* population size $4N_m N_f/(N_m + N_f)$ replacing N, see Example 2.9.

A. Etheridge, *Some Mathematical Models from Population Genetics*, Lecture Notes in Mathematics 2012, DOI 10.1007/978-3-642-16632-7_2,
© Springer-Verlag Berlin Heidelberg 2011

From this definition it is an elementary matter to work out the genealogical trees that relate individuals in a sample from the population. Suppose first that we take a sample of size two. The probability that these two individuals share a common parent in the previous generation is $1/N$. If they do not, then the probability that their parents had a common parent is $1/N$, and so on. In other words, the time to the most recent common ancestor (MRCA) of the two individuals in the sample has a geometric distribution with success probability $1/N$. (The probability that their most recent common ancestor was T generations in the past is pq^{T-1} where $p = 1/N$ and $q = 1 - p$.) In particular, the expected number of generations back to their MRCA is N. Now typically we are interested in large populations, where our rather crude models have some hope of having something meaningful to say. Then it makes sense to measure time in units of size N and in those units the time to the MRCA of a sample of size two is approximately exponentially distributed with parameter one. More generally, consider a sample of size $k \geq 2$. The probability of three (or more) individuals from the sample sharing a common parent is $\mathcal{O}(1/N^2)$ and similarly the chance that two separate pairs of individuals are 'siblings' is $\mathcal{O}(1/N^2)$. This means that the time we must wait before we see such an event is $\mathcal{O}(N^2)$ generations. But before this happens (with probability tending to one as $N \to \infty$) all our ancestral lineages will have coalesced through pairwise coalescence events (each of which occurs within $\mathcal{O}(N)$ generations). Thus the time (in units of size N) before the present at which we first see a 'merger' of lineages ancestral to our sample is approximately exponentially distributed with rate $\binom{k}{2}$ and, when that merger takes place, it affects exactly two lineages chosen uniformly at random from the $\binom{k}{2}$ pairs available. After that we just trace the remaining $\binom{k-1}{2}$ pairs of lineages and the same picture holds.

Remark 2.2. Since we are dealing with a haploid population, each individual has only one parent and the genealogical trees get *smaller* as we go backwards in time, in contrast to our usual understanding of family trees (for a diploid population) which grow as we trace backwards in time. We'll return to this point in Sect. 2.7.

We shall loosely refer to the system of coalescing lineages that we have just described as Kingman's coalescent, but let us give a more formal definition. If we label individuals in our sample $\{1, 2, \ldots, k\}$, then our process of coalescing lineages defines a continuous time Markov process, $\{\pi_t\}_{t \geq 0}$, on the equivalence relations on $[k] = \{1, 2, \ldots, k\}$. Each equivalence class of π_t corresponds to an ancestor alive at time t before the present. It consists of the labels of all individuals in our sample descended from that ancestor.

Definition 2.3 (Kingman coalescent). A k-coalescent is a continuous time Markov chain on $\mathcal{E}_k$, the space of equivalence relations on $[k]$, with transition rates $q_{\xi,\eta}$ ($\xi, \eta \in \mathcal{E}_k$) given by

$$q_{\xi,\eta} = \begin{cases} 1 & \text{if } \eta \text{ is obtained by coalescing two of the equivalence classes of } \xi, \\ 0 & \text{otherwise.} \end{cases}$$

The *Kingman coalescent* on $\mathbb{N}$ is a process of equivalence relations on $\mathbb{N}$ with the property that, for each k, its restriction to $[k]$ is a k-coalescent. By convention, we take the initial condition to be the trivial partition into singletons.

Remark 2.4 (Consistency). If we take a $(k+l)$-coalescent and restrict it to $[k]$, then we obtain a k-coalescent. In particular, if we take a sample of size $k+l$ from the population and restrict the genealogical trees relating the full sample to a randomly chosen subsample of size k, then we arrive at the same trees as if we had just taken a smaller sample in the first place. This sampling consistency is an essential part of the interpretation of the model.

Existence of the k-coalescent is clear (it is a finite state space Markov chain with bounded rates). The consistency allowed Kingman (1982) to pass to a projective limit.

Remark 2.5 (Terminology). In what follows we shall sometimes say that the genealogy of a sample (or population) of size k is determined by the Kingman coalescent. By this we mean that it is given by a k-coalescent.

To obtain the Kingman coalescent, we measured time in units of population size N and passed to an infinite population limit. Now let's examine what happens when we let $N \to \infty$ in our Wright–Fisher model. Suppose that the gene in question has two alleles which we label a and A say. For now we suppose that an offspring inherits the allelic type of its parent. We try to characterise the process, $\{p_t\}_{t \geq 0}$, which records the proportion of a-alleles in the population at each time $t \geq 0$. Notice that in the preliminary model, $\{p_t\}_{t \geq 0}$ is a discrete time Markov chain on a finite state space with traps at 0 and 1.

Definition 2.6 (Fixation). If the proportion of one of the alleles in the population is one, then we say that the allele has *fixed*. The probability that a becomes fixed is its *fixation probability*.

To characterise the distribution of $\{p_t\}_{t \geq 0}$, we consider how $\mathbb{E}[u(p_t)]$ changes with time for sufficiently nice functions $u : [0, 1] \to \mathbb{R}$. In the rescaling that we took to obtain the Kingman coalescent, the model evolves at time intervals of length $1/N$. Evidently, if a proportion p of the population is of type a in the current generation, then the expected *number* of type a individuals in the next generation is Np and the variance of that number is Npq (where $q = 1 - p$). Thus the mean allele frequency remains the same and the variance is pq/N. Moreover $\mathbb{E}[(p_{1/N} - p)^k \,|\, p_0 = p] = \mathcal{O}(1/N^2)$ for all $k \geq 3$. Now the evolution of the process is homogeneous in time, so it is enough to consider what happens close to time zero. Using Taylor's Theorem, we obtain

$$\frac{d}{dt}\mathbb{E}[u(p_t) \,|\, p_0 = p]\Big|_{t=0} \approx \frac{\{\mathbb{E}[u(p_{1/N}) \,|\, p_0 = p] - u(p)\}}{1/N}$$

$$= N\Big\{u'(p)\mathbb{E}[(p_{1/N} - p) \,|\, p_0 = p]$$

$$+ \frac{1}{2}u''(p)\mathbb{E}[(p_{1/N} - p)^2 \,|\, p_0 = p] + \mathcal{O}\left(\frac{1}{N^2}\right)\Big\}$$

$$= \frac{1}{2}p(1-p)u''(p) + \mathcal{O}\left(\frac{1}{N}\right).$$

Thus, in the limit as $N \to \infty$, if the process of allele frequencies converges to a well-defined stochastic process, then we expect that

$$\frac{d}{dt}\mathbb{E}\left[u(p_t)\vert\, p_0 = p\right]\Bigg|_{t=0} = \frac{1}{2}p(1-p)u''(p). \tag{2.1}$$

That is, we expect that in the limit, the distribution of the allele frequencies is governed by the solution to the Wright–Fisher stochastic differential equation:

$$dp_t = \sqrt{p_t(1-p_t)}dW_t, \tag{2.2}$$

where $\{W_t\}_{t\geq0}$ is a standard Brownian motion.

What we have *shown* is that, at least for large populations evolving according to the neutral Wright–Fisher model, if we measure time in units of N generations, then the distribution of allele frequencies should be approximately governed by the partial differential equation (2.1), and the genealogy of a sample from the population should be well-approximated by the Kingman coalescent. Notice that it is the random genetic drift, that is the random change in allele frequencies caused by the random variation in individual reproduction, that causes coalescence of ancestral lineages as we trace backwards in time.

In reality, a variety of factors affect the rate of genetic drift and these are often summarised by using an *effective* population size.

Definition 2.7 (Effective population size). The effective population size N_e of a population is the size of the Wright–Fisher population that would give the same rate of random drift.

Remark 2.8. In fact this definition is incomplete as there are several ways to define the rate of genetic drift and they do not necessarily yield the same expression for the effective population size. For the Wright–Fisher model for a population of size N, we have the following three properties:

1. The maximum nonunit eigenvalue of the transition matrix is $1 - 1/N$.
2. The probability that two genes taken at random are descendants of the same parent is $1/N$.
3. Writing $p(t)$ for the proportion of a-alleles in generation t and $\mathrm{var}(p(t))$ for the corresponding variance, given $p(t)$, $\mathrm{var}(p(t+1)) = p(t)(1 - p(t))/N$.

One can try to find an N_e corresponding to any of these properties, and this leads to *eigenvalue* effective population size, *inbreeding* effective population size and *variance* effective population size. Ewens (1982) discusses this in more detail. Nordborg and Krone (2002) define the *coalescent* effective size as the amount by which time must be rescaled in order to recover the Kingman coalescent as the genealogy in the limit as population size tends to infinity. Such an effective size may not exist, but there are strong arguments for not defining an effective population size in settings where one cannot (asymptotically) reduce to Kingman's coalescent. This is discussed further in Sjödin et al. (2005).

For a diploid population, modelled as a haploid population of size $2N$, the corresponding quantity will be $2N_e$.

Example 2.9 (Populations that are subdivided into males and females). Suppose that a diploid population is subdivided into N_m males and N_f females, then

$$N_e = \frac{4N_mN_f}{N_m+N_f}. \qquad (2.3)$$

To see why, take a sample of two genes from the current generation. Each sits in a diploid individual and has probability $1/2$ of being inherited from the father of that individual and $1/2$ of being inherited from the mother.[2] If they are both inherited from fathers, which happens with probability $1/4$, then they have probability $1/2N_m$ of being descended from the same gene, and similarly, if both are inherited from a female they came from the same parental gene with probability $1/2N_f$. Thus the chance of coalescence in the previous generation is

$$\frac{1}{4}\frac{1}{2N_m} + \frac{1}{4}\frac{1}{2N_f} = \frac{N_m+N_f}{8N_mN_f} = \frac{1}{2N_e}$$

with N_e given by (2.3).

What we have derived here is the inbreeding effective population size, but the methods of Sect. 6.3 can be used to show that in this example this corresponds to the coalescent effective population size (see Nordborg and Krone (2002) for more details.) □

So how does (2.2) do as a model? Of course it is too simplistic to apply to most naturally occurring populations, but we can compare it to experimental data. Buri (1956) reports an experiment on populations of *Drosophila melanogaster*. Just over one hundred populations were propagated, each from eight males and eight females. The experiment measures the frequency of an allele of a gene that slightly alters the eye colour (without affecting fitness or reproductive success of the carrier). We'll denote it by a. He reports the change in the *variance* in allele frequency across the different populations with time. All populations are started with exactly half a and half A (which in this context just means 'not a') alleles. The variance starts at zero (all populations have the same frequencies) and then grows because of the random genetic drift until it reaches a maximum when each population consists either entirely of a-alleles or entirely of A-alleles.

We write v_t for the variance in allele frequency across populations at time t in our rescaled time units, $v_t = \mathbb{E}[p_t^2] - \mathbb{E}[p_t]^2$. Using (2.1) and the Markov property of $\{p(t)\}_{t\geq0}$ we have that

$$\frac{d}{dt}\mathbb{E}[p_t] = 0, \quad \frac{d}{dt}\mathbb{E}[p_t^2] = \mathbb{E}[p_t(1-p_t)] \quad \text{and} \quad \frac{d}{dt}\mathbb{E}[p_t(1-p_t)] = -\mathbb{E}[p_t(1-p_t)].$$

---

[2] We are ignoring the possibility that we have sampled two distinct genes from the *same* individual. If this happens, then in the previous generation the ancestral lineages were necessarily in different individuals (the mother and father) and so correcting for this makes negligible difference to the inbreeding effective population size.

Combining these gives that $v_t \approx p_0(1 - p_0)(1 - \exp(-t))$. Writing V_t for the variance after t generations (in other words changing back to 'real' time units) this becomes

$$V_t \approx p_0(1 - p_0)(1 - \exp(-t/2N)).$$

The $2N$ is because Drosophila are diploid and in this case $N = 16$.[3] The theoretical prediction for the rate of increase in the variance turns out to be not very accurate, but it becomes good when instead of substituting the actual population size, one substitutes a smaller, effective, population size. Buri reports a best fit of $N_e = 11.5$. Buri's data and the theoretical predictions for $N_e = 16$ and $N_e = 11.5$ are plotted in the graph in Fig. 2.1.

Remark 2.10 (Large populations). The population size $N = 16$ does not perhaps seem particularly large. However, calculating directly with the Wright–Fisher model gives a variance after t generations of

$$p_0(1 - p_0)\left(1 - \frac{1}{N}\right)^t.$$

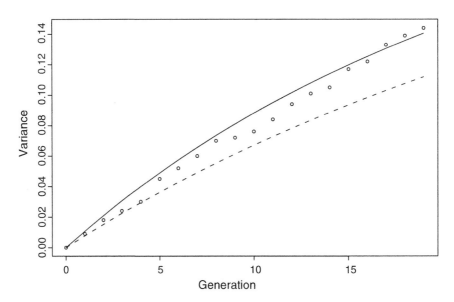

Fig. 2.1 Testing Wright's model of genetic drift. The graph shows experimental results of Buri (1956) based on just over 100 populations of *Drosophila melanogaster*, each propagated from 8 males and 8 females. Variance in allele frequency is plotted against time (in generations). Circles are data points, the dotted line is the theoretical prediction for $N_e = 16$ and the solid line is the theoretical prediction with $N_e = 11.5$

[3] Although the population was subdivided into males and females, the experiment maintained equal numbers of males and females so that the effective population size $N_e = 4N_mN_f/(N_m + N_f) = N$ (see Example 2.9).

Hence at the end of the experiment (after 19 generations) we are using $\exp(-19/32)$ as an approximation to $(1 - 1/32)^{19}$, giving a relative error of less that 1%.

But why did we need to use an effective population size here? At first sight Buri's populations appear to satisfy the assumptions of the Wright–Fisher model: they are panmictic and constant size, generation times are discrete and the allele under consideration does not affect fitness. In fact it is the Wright–Fisher reproduction mechanism itself that is at fault. It forces the variance of the offspring of a single individual to be $(1 - 1/N)$, but this does not reflect the true offspring distribution in the population. To see how offspring variance feeds into the effective population size we must consider a slightly more general model.

2.2   The Cannings Model

First a definition.

Definition 2.11 (Exchangeable random vector). A random vector $(v_1, \ldots, v_N)$ is said to be *exchangeable* if its law is invariant under any permutation of the coordinates. That is,

$$(v_1, \ldots, v_N) \overset{d}{=} (v_{\pi(1)}, \ldots, v_{\pi(N)})$$

for any permutation $\pi = (\pi(1), \ldots, \pi(N))$ of $\{1, \ldots, N\}$.

Definition 2.12 (Neutral Cannings Model). Consider a panmictic, haploid population of constant size N. Labelling the individuals in generation t by $\{1, \ldots, N\}$, in a neutral *Cannings model*, generation $t + 1$ is determined by an exchangeable random vector $(v_1(t), \ldots, v_N(t))$ with $\sum_{k=1}^{N} v_k(t) = N$. Here, $v_k(t)$ denotes the number of children of the kth individual and the vectors $\{(v_1(t), \ldots, v_N(t))\}_{t \in \mathbb{N}}$ are assumed to be independent and identically distributed.

Notice that, mathematically, neutrality is captured by exchangeability.

The Wright–Fisher model is the special case of the Cannings model in which $(v_1(t), \ldots, v_N(t))$ has the multinomial distribution with N trials and equal weights.

Let's examine the genealogy of a sample from a large population evolving according to a more general Cannings model. Let c_N denote the probability that two individuals chosen at random from some generation have a common parent in the previous generation. Then (dropping the argument t)

$$c_N = \frac{\mathbb{E}[v_1(v_1 - 1)]}{N - 1}.$$

To see this, condition on the vector $(v_1, v_2, \ldots, v_N)$ that determines the division of offspring into families. The chance that two offspring (sampled at random and *without* replacement) both fall among the v_1 individuals that make up the first family is just $v_1(v_1 - 1)/N(N - 1)$. Now average over the distribution of the vector $(v_1, v_2, \ldots, v_N)$. This gives the probability that both offspring are in the first family. Using exchangeability, the probability that they both belong to the same family (but any one of the N available) is just N times this probability, that is $\mathbb{E}[v_1(v_1-1)/(N-1)]$ as required. (For the Wright–Fisher model, $c_N = 1/N$.) The time until the MRCA of a random sample of size two from the population will be geometric with success probability c_N. This will determine the right time scaling to get convergence to a nontrivial limit as $N \to \infty$. We are going to assume that $c_N \to 0$ as $N \to \infty$. Now consider a sample of size three. The chance that they *all* have a common parent is

$$\frac{\mathbb{E}[v_1(v_1-1)(v_1-2)]}{(N-1)(N-2)}.$$

Thus, if we measure time in units of $1/c_N$, provided that

$$\frac{\mathbb{E}[v_1(v_1-1)(v_1-2)]}{N^2 c_N} \to 0 \qquad \text{as } N \to \infty, \tag{2.4}$$

in the limit as $N \to \infty$ we will only ever see pairwise mergers. In fact it turns out Möhle (2000) that the condition in (2.4) guarantees both that $c_N \to 0$ and that

$$\frac{\mathbb{E}[v_1(v-1)v_2(v_2-1)]}{N^2 c_N} \to 0 \quad \text{as } N \to \infty,$$

so that, measuring time in units of $1/c_N$, asymptotically we will not see *simultaneous* mergers of two different pairs of ancestral lineages. In the limit as $N \to \infty$ we recover Kingman's coalescent.

Lemma 2.13. *If we sample k individuals from a population evolving according to the neutral Cannings model of Definition 2.12 and if the condition (2.4) is satisfied, then for large N, when measured in time units of $1/c_N$ generations, the genealogy of the sample is approximately a k-coalescent.*

Similar calculations to those that we did for the Wright–Fisher model show that, again measuring time in units of $1/c_N$ generations and under assumption (2.4), the distribution of allele frequencies for a sufficiently large population evolving according to the Cannings model will be governed (approximately) by the partial differential equation (2.1). The only difference from the Wright–Fisher setting is that now when we wish to compare to data we must remember that c_N is (approximately) $\text{var}(v_1)/N$, where var denotes variance. In our previous language, the *effective population size* is $N_e = N/\text{var}(v_1)$. In particular, the greater the variance in offspring number, the smaller the effective population size and the faster the rate of random drift.

Remark 2.14 (Robustness of Kingman's coalescent). In passing to an infinite population limit, we aim to find an approximation that reflects the key features of our population (in this case that it is neutral, panmictic and of constant size), but which is insensitive to the fine details of the prelimiting model. As we can already see, the Kingman coalescent approximates a wide variety of local structures and it is this robustness that makes it such a powerful tool. Forwards in time we have taken a *diffusion approximation*, approximating the Wright–Fisher model by a Wright–Fisher diffusion. The importance of diffusion approximations in population genetics can be traced to the seminal work of Feller (1951).

2.3 Selfing

In footnote 1 we remarked that in considering a haploid population of size $2N$ in place of a diploid population of size N, since each individual samples its two parents independently *with* replacement, we are allowing a small probability of *self-fertilisation*. For that model, the probability of self-fertilisation is very small (for large populations), but for many plant populations a significant proportion of off-spring are produced through self-fertilisation, or *selfing*. What effect does this have on the genealogy of a sample from such a population?

We consider a population of N diploid individuals. Let us write s for the expected fraction of offspring to be produced by selfing (in which case both genes in the offspring are sampled from the *same* parent) and $1 - s$ for the expected fraction to be produced by random mating. To understand what is happening we trace the history of two ancestral lineages. At any time in the past they can be in one of three states:

1. Two lineages in distinct individuals;
2. two lineages in the same individual;
3. coalesced.

Suppose that the two lineages are in distinct individuals. They will remain in this state for a geometrically distributed number of generations with parameter $1/N$. At that time, with probability $1/2$ they are derived from the same parental chromosome and they coalesce and with probability $1/2$ they move to the second state, two lineages in the same individual. If this individual was produced by selfing, which happens with probability $s + \mathscr{O}(1/N)$,[4] then with probability $1/2$ the lineages are derived from the same parental chromosome, and so they coalesce, and with probability $1/2$ they are derived from different parental chromosomes and they remain in state 2. Thus two lineages in state 2 remain there for a geometrically distributed number of generations with parameter

$$\frac{s}{2} + 1 - s + \mathscr{O}\left(\frac{1}{N}\right)$$

at which time the lineages coalesce with probability

---

[4] The $\mathscr{O}(1/N)$ correction in these calculations is because random mating carries a small probability of selfing.

$$\frac{s/2}{s/2+(1-s)} + \mathcal{O}\left(\frac{1}{N}\right) \approx \frac{s}{2-s},$$

otherwise the system returns to the first state. In particular, we only stay in the second state for $\mathcal{O}(1)$ generations.

If we measure time in units of N generations and let $N \to \infty$, then the second state becomes instantaneous. If the process starts in this state, then it instantaneously coalesces (with probability $s/(2-s)$) or else moves to the first state. Similarly a proportion $s/(2-s)$ of transitions from the first state to the second state will be followed by an instantaneous coalescence while the rest will be followed by instantaneous return to the first state. The overall rate of transition from the first to the third state (in rescaled time) is then

$$\frac{1}{2} + \frac{1}{2}\frac{s}{2-s} = \frac{1}{2-s}.$$

To see this, the first term corresponds to the (rescaled) rate at which two lineages in distinct individuals sample the same parental chromosome, the second is the rate at which they sample different chromosomes within the same individual – that is move to the now instantaneous state 2 – multiplied by the probability that they exit state 2 through coalescence. Alternatively, by measuring time in units of $(2-s)N$, a pair of lineages waits an exponentially distributed amount of time with parameter one before coalescing.

This argument can be extended to arbitrary finite samples from the population. It is tedious because we must keep track of many possible states. The argument above is from Nordborg and Donnelly (1997). A rigorous mathematical proof (using the techniques of Sect. 6.3) can be found in Möhle (1998). We have the following result.

Lemma 2.15. *In a diploid population as above in which a portion s of offspring are produced by selfing and the remainder by random mating, as the population size N tends to infinity, the genealogy of a sample is determined by a Kingman coalescent in which each pair of lineages coalesces at rate $2N_e$ where the effective population size $N_e = \frac{2-s}{2}N$.*

2.4 Adding Mutations

A mutation is formally defined as a "heritable change in the genetic material (DNA or RNA) of an organism". Mutations occur in many forms, but for simplicity we concentrate on *point mutations* which occur when there is a change from one base pair to another at a single position in the DNA sequence. Because of the redundancy in the genetic code some point mutations do not lead to a change in the sequence of amino acids. These are called *synonymous mutations*. Mutations are the ultimate source of all genetic variation; without them there would be no evolution. Although mutation rates are relatively slow, the mixing of mutations from different lineages that results from genetic recombination (see Sect. 5.6) rapidly leads to an enormous number of combinations on which natural selection can act. Mutation rates vary according to the type of mutation, the location on the genome and the organism involved, with the highest rates being in viruses.[5]

Typically in our models we assume a constant probability μ per individual per generation of a mutation at a given base or more generally at a given locus. If we follow a particular ancestral lineage in our population, then we must wait a geometrically distributed number of generations (with mean $1/\mu$) until we see a mutation. Assuming that $2N_e\mu$, that is the mutation rate multiplied by the effective population size, is of order one, this will, in rescaled time, be approximately exponential. Moreover, under this condition, the probability that we see both a coalescence and a mutation in our sample in a single generation is $\mathcal{O}(1/N_e^2)$. So just as in our derivation of the Kingman coalescent, we see that if there are currently k lineages ancestral to our sample, the time (in rescaled units) we must trace back until we see *some* event is (approximately) the minimum of k independent exponential random variables each with parameter $2N_e\mu$ and an independent exponential random variable with parameter $\binom{k}{2}$. Another way to say this is we can add mutations to Kingman's coalescent by simply superposing a Poisson process of mutations on the ancestral lineages. Notice that in order to ensure that the types in the sample are consistent with the pattern of mutations stemming from such a Poisson process, a type must first be assigned to the MRCA and then we work our way back through the coalescent tree assigning types to ancestral lineages. This is illustrated by example in Fig. 2.2. There are several important models of mutation. Perhaps the simplest is the parent-independent mutation model.

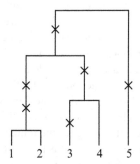

Fig. 2.2 Adding mutations to the Kingman coalescent. Mutations are added to the Kingman coalescent by throwing down an independent Poisson process of mutations on each branch. In order to ensure that the types in the sample are consistent with the pattern of mutations, one must first assign a type to the MRCA and then work back through the tree. In this example, we have used '×' to denote a mutation on a branch of the coalescent tree. Notice that here the individuals labelled 1 and 2 in the sample must have the same type

---

[5] See Fig. 12.23 in Barton et al. (2007). Rates shown there vary from $\mathcal{O}(10^{-4})$ per base pair per generation in RNA viruses like HIV to $\mathcal{O}(10^{-10})$ or $\mathcal{O}(10^{-11})$ in organisms like humans and mice. By contrast, there is a relatively uniform rate of mutation *per genome* per replication across diverse organisms.

Definition 2.16 (Parent independent mutation). In the *parent-independent* mutation model, a gene is assumed to occur in one of a finite number of types. Mutations occur at a constant rate per individual, independent of the current type of the individual. The type created by the mutation event is chosen according to a probability distribution which is also independent of the type of the parent.

More generally one can allow the probability of mutation to different types to depend on the current state of an individual, in which case the type of a lineage is governed by a Markov chain on the space of possible types.

Definition 2.17 (Infinitely many alleles model). In the *infinitely many alleles* model, every time a mutation occurs, it is to a new allele, never seen before in the population.

The infinitely many alleles model can be seen as the limit of the parent-independent mutation model when the number of alleles tends to infinity. It is useful in providing a link between the classical notion of *probability of identity* and the coalescent. In the infinitely many alleles model, two genes will be identical (that is they will have the same allelic state) if there has been no mutation since their MRCA. If their MRCA occurred T generations in the past, and the mutation rate per individual per generation is μ, then we see that this has probability $(1 - \mu)^{2T} \approx e^{-2\tilde{\mu}\tau}$, where $\tau = T/N_e$ is the time to the MRCA in the coalescent timescale and $\tilde{\mu} = N_e\mu$ is the scaled mutation rate. Averaging out over the distribution of τ, the probability of identity is $\mathbb{E}[\exp(-2\tilde{\mu}\tau)]$, that is the Laplace transform of the distribution of the time, τ, to the MRCA.

Remark 2.18 (Mutation rates and nucleotide diversity). Since the expected number of generations since the MRCA of two genes sampled at random from a diploid population (under Kingman's coalescent) is $2N_e$, on average we expect them to differ by $4N_e\mu$ mutations per base pair. This can be counted directly if we are dealing with DNA sequences. The proportion of nucleotides that differ between two randomly chosen sequences is called the *nucleotide diversity* and is usually denoted by π. The crucial parameter $4N_e\mu$ is denoted by θ. Notice then that if we measure time in units of $2N_e$ generations (as is usual for the Kingman coalescent for a diploid population) then the rate at which we see mutations falling on each ancestral lineage is $\theta/2$. This explains the choice of scaling for the mutation rate in much of what follows. We shall use the same notation when μ is no longer the mutation rate per base pair, but rather the mutation rate for a locus or a whole gene.

If we sample a single nucleotide at random then with high probability all individuals in our sample will be identical. (A locus is usually defined to be polymorphic if the frequency of the most common type is less than 0.99. In humans, the chance of *heterozygosity* at a randomly chosen nucleotide is about 0.0008. In *Drosophila* it is an order of magnitude bigger, but still only about 1%, Lynch and Conery (2003), Fig. 1.) If the rate of mutation does not vary too greatly between bases then this justifies the so-called *infinitely many sites* model in which each time we see a polymorphic site in our sample we assume that it is due to a unique mutation.

Definition 2.19 (Infinitely many sites model). In the *infinitely many sites* model, every time a mutation occurs on a lineage it is at a new position on the DNA sequence.

It is sometimes convenient to model the genome as continuous, for example as $[0, 1]$, in which case we suppose that each new mutation occurs at a position chosen according to an independent uniformly distributed random variable on $[0, 1]$.

Notice that whereas in the infinitely many alleles model individuals only carry information about the most recent mutation on their ancestral lineage, in the infinitely many sites model they retain information about *all* mutations experienced by their ancestors.

2.5 Inferring Genealogies From Data

The genealogy of a sample from a population contains a great deal of information, but we cannot observe it directly. Instead we try to infer it from the pattern of mutations in the sample. We assume the infinitely many sites mutation model. Once a mutation occurs, it will be carried by all descendants of that individual and from this we can reconstruct at least partial information about the genealogical trees. If we suppose, for simplicity, that we know which is the ancestral type at each locus, then we can construct the so-called *gene tree*. The gene tree has mutations as its vertices. Figure 2.3 shows how this works in an example. Although a given pattern of mutations may be consistent with several different coalescent trees, if it is compatible with this model then it will be consistent with an essentially unique gene tree. The gene tree is unique up to permutations of labels along single lineages (for example 1 and 2 in the example in Fig. 2.3). However, there may be *many* different corresponding coalescent trees with mutation. For example, the gene tree in Fig. 2.3 is compatible with the coalescent tree of Fig. 2.4. It is also compatible with the coalescent in which b and c coalesce before a and b. More generally, if there are insufficient mutations then coalescent trees with many different topologies

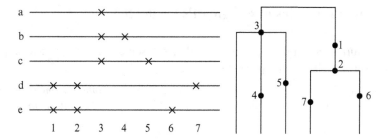

Fig. 2.3 Reconstructing a gene tree. The picture on the left represents a possible pattern of mutations in a sample of size 5. We suppose for simplicity that we know which is the ancestral type at each locus, so that an '×' in the picture indicates that an individual carries a mutation at that locus. On the right is a gene tree compatible with this pattern

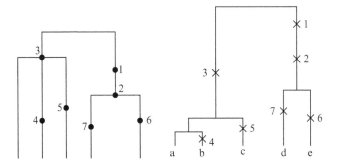

Fig. 2.4 Gene trees and coalescent trees. A given gene tree may be compatible with more than one coalescent tree. The coalescent tree on the right is compatible with the gene tree on the left. It would also be compatible with one in which b and c coalesce before a and b

may be compatible with a gene tree. (As an extreme example suppose that there were just one mutation shared by all but one individual in the sample.)

There are simple conditions to check that data is compatible with this model and efficient algorithms for reconstructing the gene trees. If the ancestral type is not known, then an unrooted tree is constructed. To recover a rooted tree one can compare to a more distantly related sequence (called an *outgroup*).

This procedure tells us something about the shape of the genealogical tree, but nothing about the lengths of the edges. However, since mutations are assumed to fall at an (approximately) exponential rate, some information about the time represented by an edge is available from the number of mutations occurring there. For much more on ancestral inference from gene trees we refer to Griffiths (2002). In practice, of course, things are not quite this simple. There are two principal problems. The first is *convergence*: if a site is evolving quickly, or if two sequences in our sample are very distantly related, then the same mutation may occur twice. The second is *recombination*, which we'll describe in more detail in Sect. 5.6. The result of recombination is that different stretches of our DNA sequence have different genealogies.

2.6 Some Properties of Kingman's Coalescent

We now return to Kingman's coalescent and record some of its elementary properties (and some of their consequences).

Lemma 2.20. *Let W_k denote the time to the most recent common ancestor of a sample of k genes whose genealogy is determined by Kingman's coalescent. Then*

$$\mathbb{E}[W_k] = 2\left(1 - \frac{1}{k}\right).$$

Proof. Since $W_k = T_k + T_{k-1} + \cdots + T_2$ where T_i is exponentially distributed with rate $\binom{i}{2}$ we have

$$\mathbb{E}[W_k] = \sum_{i=2}^{k} \frac{2}{i(i-1)}$$

$$= 2 \sum_{i=2}^{k} \left[\frac{1}{i-1} - \frac{1}{i} \right]$$

$$= 2 \left(1 - \frac{1}{k} \right).$$

□

Thus the mean time to the MRCA of the whole population (k infinite) is only twice that for a sample of size two. The picture is that for a large sample, as we trace backwards in time, we see a burst of quick coalescence followed by a long period with just a few ancestors. As a result, adding more and more individuals to our sample adds surprisingly little information. Moreover, since, in 'real' time, the standard deviation of the time when there are exactly two ancestral lineages is N_e generations (or twice that for a diploid population), the tree is always highly variable irrespective of the sample size. Figure 2.5 is a simulation of the Kingman coalescent for a sample of size 1,000.

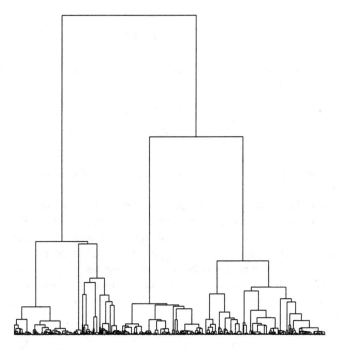

Fig. 2.5 Simulation of the Kingman coalescent. The picture is a single realisation in a simulation (courtesy of Bob Griffiths) of a Kingman coalescent for a sample of size 1,000. Notice the initial period of very rapid coalescence. For a large proportion of the time back to the MRCA, only two or three ancestral lineages remain

Lemma 2.21. *Let $L^{(k)}$ denote the total length of the genealogical tree relating a sample of size k. Under the Kingman coalescent, $L^{(k)}/2$ is distributed as the maximum of $(k-1)$ independent exponential random variables. In particular,*

$$\frac{1}{2}L^{(k)} - \log k \xrightarrow{d} X \qquad as\ k \to \infty,$$

where X has a Gumbel distribution with density $\exp(-x - e^{-x})$.

Proof. The length of the tree here is measured until the time of the MRCA of the sample. Notice that if E is an exponentially distributed random variable with parameter one, then for $\gamma > 0$, writing $X_\gamma = E/\gamma$, we have $\mathbb{P}[X_\gamma > t] = \mathbb{P}[E > \gamma t] = \exp(-\gamma t)$, so that X_γ is exponentially distributed with parameter γ.

Now, in this notation, for each $2 \le j \le k$, the portion of $L^{(k)}$ corresponding to the time when there are exactly j ancestral lineages is $jX_{\binom{j}{2}}$ and the random variables $X_{\binom{j}{2}}$ are independent for different j. Thus

$$L^{(k)} = \sum_{j=2}^{k} jX_{\binom{j}{2}} = \sum_{j=2}^{k} \frac{j}{\binom{j}{2}} E_j$$

$$= \sum_{j=2}^{k} \frac{2}{j-1} E_j,$$

where the E_j are independent exponentially distributed random variables with parameter one. From this

$$\frac{1}{2}L^{(k)} = \sum_{i=1}^{k-1} \frac{1}{i} E_{i+1} = \sum_{i=1}^{k-1} X_i = \sum_{j=1}^{k-2} X_{k-j-1}, \tag{2.5}$$

where the random variables X_i are independent exponential random variables with parameter i.

Now suppose that we have $k-1$ independent exponential random variables, each with parameter one, and arrange them in increasing order, $E^{(1)} < E^{(2)} < \cdots < E^{(k-1)}$. Then $E^{(1)}$ has an exponential distribution with parameter $(k-1)$ and, as a result of the lack of memory property of the exponential distribution, for $1 \le j \le k-2$, $E^{(j+1)} - E^{(j)}$ has an exponential distribution with parameter $k-j-1$. Thus the right hand side of (2.5) is distributed exactly as the maximum of $k-1$ independent exponentially distributed random variables, each with parameter one.

In particular,

$$\mathbb{P}\left[\frac{1}{2}L^{(k)} < x\right] = (\mathbb{P}[E_1 < x])^{k-1} = \left(1 - e^{-x}\right)^{k-1},$$

and so

$$\mathbb{P}\left[\frac{1}{2}L^{(k)} - \log k < x\right] = \left(1 - e^{-(x+\log k)}\right)^{k-1} \qquad \text{for } x > -\log k$$

$$= \left(1 - \frac{1}{k}e^{-x}\right)^{k-1}$$

$$\to \exp(-e^{-x}) \qquad \text{as } k \to \infty.$$

□

Remark 2.22. Although $L^{(k)}$ has mean $2\log k$, the variance, $\text{var}(L^{(k)})$, is bounded as $k \to \infty$.

Conditional on $L^{(k)}$, under the infinitely many sites model, the number of mutations that we see in our sample is Poisson with parameter $\theta L^{(k)}/2$ (recall Remark 2.18). Each site at which we see a mutation is called a *segregating site* or *SNP* (single nucleotide polymorphism). Writing $S^{(k)}$ for the number of segregating sites, we see that

$$\frac{2S^{(k)} - \theta L^{(k)}}{\sqrt{2\theta L^{(k)}}}$$

is asymptotically normally distributed with mean zero and variance one. Thus if we know the asymptotic distribution of $L^{(k)}$ we can deduce the asymptotic distribution of $S^{(k)}$.

Definition 2.23 (Watterson's estimator). Watterson proposed the following estimator for the mutation rate:

$$\hat{\theta} = \frac{2S^{(k)}}{\mathbb{E}[L^{(k)}]} = \frac{S^{(k)}}{\sum_{i=1}^{k-1}\frac{1}{i}}.$$

As a result of Lemma 2.21 we see that Watterson's estimator is asymptotically normal. However, since $L^{(k)}$ grows like $\log k$, in practice the convergence is extremely slow.

2.7 Genealogies and Pedigrees

We have seen that under our neutral population models, in finite time everyone in our population traces back to a single common ancestor. It follows immediately (by symmetry) that if an allele starts with frequency p_0 in the population, and there is no mutation, then the probability that it is eventually fixed (that is, carried by everyone) is just p_0. As a special case, the probability that a particular gene present in a single individual now will leave descendants in the indefinite future is $1/N$.

On the other hand, if we trace back family trees in a diploid population, then each individual has two parents, four grandparents and so on and, in a finite population, we quickly exhaust the population. Of course, in practice the ancestors are not all unique, but nonetheless we expect a significant proportion of the population to be included somewhere in our family tree. We shall refer to this family tree as the *pedigree* of the individual.

The following lemma illustrates the fact that if we trace far enough back in time, *most* individuals in the ancestral population will be in the pedigree of a given individual now.

Lemma 2.24. *Suppose that in a large diploid (but for simplicity hermaphrodite) population of size N, evolving in discrete generations, each individual chooses* two *parents uniformly at random from the previous generation. Then the probability that a randomly chosen individual from the population t generations in the past is in the pedigree of a given individual in the current population converges to about 0.8 as* $t \to \infty$.

Idea of Proof. First note that since N is large, the random number of descendants left by a single individual is approximately Poisson with parameter two (being, if we ignore the possibility of an individual choosing the same parent twice, Binomial with $2N$ trials and success probability $1/N$). Let $P(t)$ be the probability that an individual alive t generations ago does *not* belong to the pedigree of our chosen individual. Then, since none of that individual's descendants can be in the pedigree, we have $P(t+1) \approx \exp(-2 + 2P(t))$.[6]

The equation $p = \exp(-2 + 2p)$ can be solved (at least numerically). To see this, we first rearrange to obtain $(-2p)\exp(-2p) = -2\exp(-2)$. Now the equation $z = W(z)\exp(W(z))$ defines the *Lambert W function*, also known as the *product log* function. In general it is multivalued, but for $z \in (-1/e, 0)$ there are just two branches and choosing the one with $W(z) \geq -1$ gives a unique solution. This yields $p = -W(-2e^{-2})/2$ which is close to 0.2. $\square$

The same calculation tells us that the 80% of individuals that are in the pedigree of our chosen individual are actually in the pedigree of *everyone* in the current population. The conclusion is that although most of us will have descendants alive into the indefinite future, a particular gene is highly unlikely to be transmitted.

In fact much finer results than these are known. Chang (1999) shows that if we go back $\sim \log_2 N$ generations[7] then we can expect to see an individual in the population who is ancestral to *every* present-day individual. Tracing back $\sim 1.77 \log_2 N$ generations *all* those individuals who are ancestors will be ancestors of every present-day individual.

---

[6] Here we are supposing that the probability of being in the pedigree is independent for each of the Poisson number of individuals. Although not *quite* true, the idea is that this probability is determined while the family trees of descendants of the different individuals are still small, before the dependence becomes important. We refer to Chang (1999) for a rigorous proof.

[7] We are using the notation $f(N) \sim g(N)$ to mean $f(N)/g(N) \to 1$ as $N \to \infty$.

Remark 2.25. In Baird et al. (2003) a branching process model is considered which traces the pedigree descendants of an individual *forwards* in time in a diploid population and asks whether that individual contributes *any* genetic material to the population t generations into the future. The genome is represented by the interval $[0, 1]$. As a result of *recombination* (see Sect. 5.6), each offspring inherits, with equal probability, either the block $[0, U]$ or the block $[U, 1]$ of genome from the 'pedigree parent', with the complement coming from the other parent (assumed unrelated). The random variable U is uniformly distributed on $[0, 1]$ and is independent for each offspring. Whereas the probability of transmission of a particular gene in such a branching process model is $\mathcal{O}(1/t)$ (corresponding to the probability that a critical branching process survives until time t) if one asks whether *some* material from a block of genome has been transmitted, the rate of decay of survival probability is much slower (of $\mathcal{O}(1/\log t)$). This effect is akin to the birthday problem, since we are just asking that *some* block be transmitted, we are not specifying a particular block.

2.8 The Moran Model

We now return to the main theme of this chapter, random genetic drift, and introduce a second important model, the *Moran model* (due to Moran (1958)). Although less popular with biologists than the Wright–Fisher model, mathematically it is often more convenient. For example, in a population divided into two allelic types (as in Sect. 2.1), the frequency of the a-allele is governed by a birth and death process which greatly simplifies its analysis. Moreover, as we shall see, the genealogy of a sample from a population evolving according to a Moran model is *exactly* determined by Kingman's coalescent.

There are two essential differences between the Wright–Fisher model and the Moran model:

1. Whereas the Wright–Fisher model evolves in discrete generations, in the Moran model generations overlap.
2. In the Wright–Fisher model an individual can have up to N offspring, but in the Moran model an individual always has zero or two offspring.

Definition 2.26 (The neutral Moran model). A population of N genes evolves according to the *Moran model* if at exponential rate $\binom{N}{2}$ a pair of genes is sampled uniformly at random from the population, one dies and the other splits in two.

Remark 2.27. There is no agreement in the literature as to how to choose the rate at which pairs of individuals are chosen, this choice is convenient as it means that the genealogy of the population is determined by Kingman's coalescent, with no need for a further time change. With this choice of parameters, therefore, we can compare the predictions of the Moran model to those of the Wright–Fisher or Cannings models in the *coalescent* timescale. However, some care is needed in interpreting the model in 'real' time units.

Remark 2.28. The embedded discrete time Markov jump chain is a Cannings model in which the vector $(v_1(t),\dots,v_N(t))$ is uniformly distributed on all the permutations of $(2,0,1,1,\dots,1)$.

A more formal way to describe the model is as follows. We suppose that individuals in our population at time zero are labelled $1,\dots,N$. Associated to each pair of labels (i,j) is an independent rate one Poisson process that we denote by $\pi_{(i,j)}$. Since there are only a finite number of these, the points of distinct $\pi_{(i,j)}$'s are distinct. At a point of the Poisson process $\pi_{(i,j)}$, the individuals (genes) currently labelled (i,j) are involved in a reproduction event in which one dies and the other reproduces (with equal probabilities). The two offspring adopt the labels i and j. This is represented graphically in Fig. 2.6.

We can recover the ancestry of a sample by tracing backwards in time. If an ancestral line is at the tip of an arrow, then it *coalesces* with that at the root. If it is at the root it will be unaffected. For the population of Fig. 2.6 this is illustrated in Fig. 2.7. It is not hard to convince oneself that the genealogical trees relating individuals in a random sample are then precisely those generated by Kingman's coalescent. For example, follow a sample of size two backwards in time. The labels of the two individuals will change with time, let's call them $(i(t),j(t))$ say, but because of the lack of memory property of the exponential distribution, the time until we see an arrow joining the pair $(i(t),j(t))$ is still going to be exponential parameter one; if a label changes before coalescence, we simply piece together the random time before the label change with the remaining random time after the label change until we see coalescence. In particular then we see that, for large populations, from the point of view of the genealogy of a sample it makes little difference whether we consider a Wright–Fisher model or a Moran model.

Remark 2.29 (Adding mutations). We should like to add mutations to the Moran model in such a way that we can readily make comparisons with the Wright–Fisher model. For this reason, we separate the processes of mutation and reproduction so

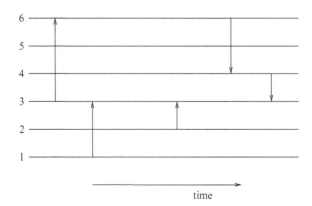

time

Fig. 2.6 Graphical representation of the Moran model. We draw an arrow between the lines labelled (i,j) at each point of $\pi_{(i,j)}$. The arrow $i \to j$ indicates that i reproduced and j died, $i \leftarrow j$ indicates that j reproduced and i died

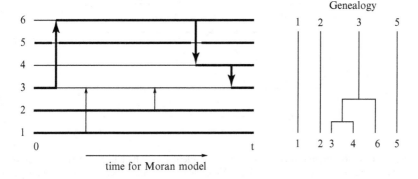

Fig. 2.7 Genealogies under the Moran model. The picture on the right shows the genealogical trees relating individuals in the population on the left, tracing back from time t to time 0.

that mutations fall on the genealogical tree relating individuals in the sample according to a Poisson process, just as in Sect. 2.4. Since we are already in the timescale of the Kingman coalescent (c.f. Remark 2.27), it is natural to suppose that each individual accumulates mutations at a constant rate (irrespective of population size). In order to incorporate a range of different mutation models, we model this by supposing that in between reproduction events, the type of each individual, independently, evolves according to a mutation process (typically, but not necessarily, a finite state space Markov chain).

2.9 The Site Frequency Spectrum

In this section we exploit the relationship with the Moran model to continue our investigation of the Kingman coalescent.

The simplest statistic for a sample under the infinitely many sites mutation model is the number of segregating sites, whose distribution we discussed in Sect. 2.6, but one can also ask for more detailed information.

Definition 2.30 (Site frequency spectrum). For a sample of size k under the infinitely many sites mutation model, write $M_j(k)$ for the number of sites at which exactly j individuals carry a mutation. The vector $(M_1(k), M_2(k), \ldots, M_k(k))$ is called the *site frequency spectrum* of the sample.

This is illustrated in Fig. 2.8.

Lemma 2.31. *Suppose that the genealogy of a sample is determined by the Kingman coalescent and that mutations occur at rate $\theta/2$ along each ancestral lineage. Under the infinitely many sites mutation model we have*

$$\mathbb{E}[M_j(k)] = \frac{\theta}{j}. \tag{2.6}$$

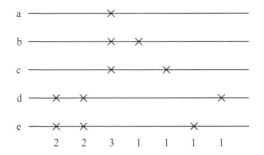

Fig. 2.8 The site frequency spectrum. In the example depicted here there are seven mutations, four of which are singletons, one occurs three times and two appear twice. (We are assuming that we know the ancestral type at each locus.) The site frequency spectrum is $(4,2,1,0,0)$

Proof of Lemma 2.31. We use the relationship between the Kingman coalescent and the Moran model. We emphasise that the 'population' in the Moran model below is not that from which we have sampled. It will have size k, the number of individuals in the *sample*. Suppose that a mutation arose at time $-t$ (that is t before the present) and denote individuals in our sample carrying that mutation as type a. For the corresponding Moran model (with population size k), we think of the mutation as arising at time zero and of the sample as the *whole population* at time t.

From the point of view of the Moran model, the probability that we see j type a individuals in the sample is the probability that a mutation arising on a single individual at time zero is carried by j individuals at time t later. We write X_t for the number of type a individuals at time t and $p(t,i,j) = \mathbb{P}[X_t = j \mid X_0 = i]$. In this notation, the probability that there are exactly j type a individuals in the sample is $p(t,1,j)$.

Since, under the infinitely many sites model, each mutation occurs at a different point on the genome and mutations occur at rate $\theta/2$ per individual (and the population size is k), the expected *total* number of sites at which we see a mutation carried by exactly j individuals is just

$$\mathbb{E}[M_j(k)] = \int_0^\infty k \frac{\theta}{2} p(t,1,j) dt. \tag{2.7}$$

Now $G(i,j) \equiv \int_0^\infty p(t,i,j) dt$ is just the expected total time that the process $\{X_t\}_{t \geq 0}$ spends in site j if it started from i and our next task is to calculate this.

Note that if $X_s = i$, then it moves to a new value at rate $i(k-i)$ (which is just the number of the $\binom{k}{2}$ ways of sampling a pair from the population in which the two individuals sampled are of different types) and when it does move, it is equally likely to move to $i-1$ or $i+1$. Let

$$T_i = \inf\{t > 0 : X_t = i\}$$

denote the first hitting time of site i. Then since 0 is a trap for the process we have

$$G(1,j) = \mathbb{P}[T_j < T_0 \mid X_0 = 1] \cdot G(j,j).$$

Now, because it is just a timechange of a simple random walk, for $0 \le i \le j$,

$$\mathbb{P}[T_0 < T_j \,|\, X_0 = i] = \frac{j-i}{j},$$

and similarly, for $j \le l \le k$,

$$\mathbb{P}[T_k < T_j \,|\, X_0 = l] = \frac{l-j}{k-j}.$$

Thus, partitioning on whether the first jump out of j is to $j-1$ or to $j+1$, we find that if it is currently at j, the probability that this is the *last* visit that X_t makes to j is

$$\rho = \frac{1}{2}\frac{1}{j} + \frac{1}{2}\frac{1}{k-j} = \frac{1}{2}\frac{k}{j(k-j)}.$$

In other words, if we start from j, the number of visits to j (including the current one) before either the allele is fixed in the population or it is lost is geometric with parameter ρ. Each visit lasts an exponentially distributed time with mean $\frac{1}{j(k-j)}$. Thus

$$G(1,j) = \frac{1}{j}G(j,j) = \frac{1}{j}\frac{1}{\rho}\frac{1}{j(k-j)} = \frac{2}{kj}.$$

Substituting into (2.7) completes the proof. □

Remark 2.32. The remarkable fact about this result is that the site frequency spectrum is almost independent of k. Increasing k only changes the allowed classes. The sceptical reader can work directly and, by conditioning on the first event as one traces backwards in time in the Kingman coalescent, check that the expected number of singletons is independent of k. This approach rapidly becomes tedious when checking the corresponding result for the other terms in the spectrum.

2.10 The Lookdown Process

The consistency of the k-coalescents for different values of $k \in \mathbb{N}$ allows us to recover all of them as projections of a single stochastic process, Kingman's coalescent. Since genealogical trees for the Moran model are precisely governed by the Kingman coalescent, it is reasonable to hope that we can also construct Moran models corresponding to different population sizes as projections of a single stochastic process. This is at the heart of the powerful Donnelly and Kurtz *lookdown process*.

To see how it works, we exploit the connection with the Kingman coalescent. Suppose that the population at the present time is labelled $\{1,2,\ldots,N\}$. Recall that the full description of the Kingman coalescent (or rather the N-coalescent) is as a process taking values among the set of equivalence relations on $\{1,2,\ldots,N\}$, with each ancestral lineage corresponding to a single equivalence class. Now suppose that we label each equivalence class by its smallest element. If blocks with labels $i < j$ coalesce, then after the coalescence the new block is necessarily labelled i. In our graphical representation of the Moran model, this just dictates the direction of the arrow corresponding to that coalescence event; it will always be the individual with the smaller label that gave birth. Backwards in time, our process is equivalent to one in which, as before, at the points of a rate one Poisson process $\pi_{(i,j)}$ arrows are drawn joining the labels i and j, but now the arrows are always in the same direction (upwards with our convention). The genealogies are still determined by the Kingman coalescent, we have simply chosen a convenient labelling, and so in particular they are precisely those of the Moran model. But what about forwards in time? What we saw backwards in time was that choosing the direction of the arrows corresponded to choosing a particular labelling of the population. If the distribution of the population is *exchangeable*, that is it doesn't depend on the labelling, then forwards in time too we should not have changed the distribution in our population. Our next task will be to check this, but first we need a formal definition.

Definition 2.33 (The N-particle lookdown process). The N-particle lookdown process will be denoted by the vector $(\zeta_1(t), \ldots, \zeta_N(t))$. Each index is thought of as representing a 'level', with $\zeta_i(t)$ denoting the allelic type of the individual at level i at time t. The evolution of the process is described as follows. The individual at level k is equipped with an exponential clock with rate $(k-1)$, independent of all other individuals. At the times determined by the corresponding Poisson process it selects a level uniformly at random from $\{1, 2, \ldots, k-1\}$ and adopts the current type of the individual at that level. The levels of the individuals involved in the event do not change. In between lookdown events the type at each level evolves, independently, according to the mutation process.

Remark 2.34. Because of our convention over the interpretation of arrows, it is not at all clear from the above why one should call this the *lookdown* process. The explanation is that at rate $(k-1)$ the kth individual 'looks down' to a level chosen uniformly at random from those below and adopts the type of the individual at that level.

To see that the lookdown process and the Moran model produce the same distribution of types in the population, provided we start from an exchangeable initial condition, we examine their infinitesimal generators. Recall the definition of the generator of a continuous time Markov process.

Definition 2.35 (Generator of a continuous time Markov process). Let $\{X_t\}_{t \geq 0}$ be a real-valued continuous time Markov process. For simplicity suppose that it is time homogeneous. For a function $f : \mathbb{R} \to \mathbb{R}$ define

$$\mathscr{L}f(x) = \lim_{\delta t \downarrow 0} \frac{\mathbb{E}[f(X_{\delta t}) - f(x)|X_0 = x]}{\delta t}$$

if the limit exists. We'll call the set $\mathscr{D}(\mathscr{L})$ of functions for which the limit exists the *domain* of $\mathscr{L}$, and the operator $\mathscr{L}$, acting on $\mathscr{D}(\mathscr{L})$, the *infinitesimal generator* of $\{X_t\}_{t\geq 0}$.

If we know $\mathscr{L}$, then we can write down a differential equation for the way that $\mathbb{E}[f(X_t)]$ evolves with time. If $\mathscr{L}f$ is defined for sufficiently many different functions then this completely characterises the finite dimensional distributions of $\{X_t\}_{t\geq 0}$.

Let us write E for the space of possible allelic types for individuals in our population. The Moran model for a population of size N is then simply a continuous time Markov chain on E^N and its infinitesimal generator, K_N, evaluated on a function $f : E^N \to \mathbb{R}$, is given by

$$K_N f(x_1, x_2, \ldots, x_N) = \sum_{i=1}^{N} A_i f(x_1, x_2, \ldots, x_N)$$
$$+ \frac{1}{2} \sum_{i=1}^{N} \sum_{j=1}^{N} [\Phi_{ij} f(x_1, \ldots, x_N) - f(x_1, \ldots, x_N)], \quad (2.8)$$

where $\Phi_{ij} f(x_1, \ldots, x_N)$ is the function obtained from f by replacing x_j by x_i. The operator A_i is the generator of the mutation process, A, acting on the ith coordinate. (Recall that in the Moran model mutation was superposed as a Markov process along lineages.)

The generator of the N-particle lookdown process, L_N is given by

$$L_N f(x_1, x_2, \ldots, x_N) = \sum_{i=1}^{N} A_i f(x_1, x_2, \ldots, x_N)$$
$$+ \sum_{1 \leq i < j \leq N} [\Phi_{ij} f(x_1, x_2 \ldots, x_N) - f(x_1, x_2 \ldots, x_N)]. \quad (2.9)$$

Assuming that we start both processes from the same exchangeable initial condition, we should like to show that the types $(\zeta_1(t), \zeta_2(t), \ldots, \zeta_N(t))$ under the lookdown model and the types $(Z_1(t), Z_2(t), \ldots, Z_N(t))$, say, under the original Moran process have the same distribution for each fixed $t > 0$, even though the processes are manifestly different. Following Dawson (1993), we must check that the generators of the two processes agree on symmetric functions. Observe first that any symmetric function, f, satisfies

$$f(x_1, x_2, \ldots, x_N) = \frac{1}{N!} \sum_{\pi} f(x_{\pi(1)}, x_{\pi(2)}, \ldots, x_{\pi(N)}),$$

where the sum is over all permutations of $\{1, 2, \ldots, N\}$. Substituting this expression for f into (2.9), we recover (2.8). In other words, the generators of $(\zeta_1, \zeta_2, \ldots, \zeta_N)$ and $(Z_1, Z_2, \ldots, Z_N)$ agree on symmetric functions as required. (We are implicitly assuming uniqueness of the distribution on symmetric functions corresponding to this generator. It follows from duality with the N-coalescent, but we don't allow that to detain us here.)

The key observation now is that our Nth lookdown process is simply the first N levels of the $(N+k)$th lookdown process for any $k \geq 1$. The *infinite* lookdown process can then be constructed as a projective limit.

Theorem 2.36 (Donnelly and Kurtz 1996). *There is an infinite exchangeable particle system $\{W_i, i \in \mathbb{N}\}$ such that for each N,*

$$(W_1, W_2, \ldots, W_N) \stackrel{\mathscr{D}}{=} (\zeta_1, \zeta_2, \ldots, \zeta_N),$$

where $\zeta_1, \zeta_2, \ldots, \zeta_N$ is the N-particle lookdown process.

Remark 2.37. In fact more is true. It is known that the sequence of empirical measures $\frac{1}{N} \sum_{i=1}^{N} \delta_{Z_i(t)}$ converges to a *Fleming–Viot superprocess* as $N \to \infty$. Donnelly and Kurtz also show that

$$Y = \lim_{N \to \infty} \frac{1}{N} \sum_{i=1}^{N} \delta_{W_i},$$

is a Fleming–Viot superprocess. A rapid introduction to Fleming–Viot superprocesses and further references can be found, for example, in Etheridge (2000). Rather than introduce the general Fleming–Viot superprocess, which takes its values among probability measures on the type space E, in Sect. 2.11 we shall consider what this limit looks like in the special case when E is a two-point set representing two alleles a and A, in which case it is enough to specify the evolution of the proportion of type a individuals in the population.

Since the genealogy of a sample of size k from the Moran model is a k-coalescent, and since we've seen that the genealogy of the first k levels in the lookdown process is also a k-coalescent, with this labelling we have a nice consistent way of sampling from a Moran model of arbitrary size. The genealogy of the sample is that of the first k levels in the lookdown process. The evolution of those levels does not depend on the population size – because we only ever look 'down' we don't see the population size N at all.

2.11 A More Simplistic Limit

Instead of discussing general Fleming–Viot superprocesses (which would allow us to consider essentially arbitrary type spaces) we now turn to identifying the limiting model for allele frequencies when our population is subdivided into just two types which, as usual, we label a and A. Just as in our discussion of the rescaled Wright–Fisher model, we consider the proportion, p_t, of individuals of type a at time t.

The only possible mutations are between the two types. We suppose that each type a individual mutates to type A at rate v_1 and each type A individual mutates to type a at rate v_2. Recall that for the Moran model we are already in the timescale of the Kingman coalescent and so we should think of $v_i = N\mu_i$ where μ_1 and μ_2 are the true mutation rates.

Remark 2.38. The idea that we can mutate backwards and forwards between types may seem at odds with our discussion of mutations in Sect. 2.4. Models of this type were introduced long before biologists knew about, and had access to, DNA sequences. Classically one might imagine a small number of alleles defined through phenotype, for example colour. In modern terms one can justify the model by pooling sequences into classes according to the corresponding phenotype.

The generator for the birth-death process of allele frequencies under the Moran model for a population of size N is then

$$\mathscr{L}_N f(p) = \binom{N}{2} p(1-p) \left(f\left(p + \frac{1}{N}\right) - f(p) \right)$$

$$+ \binom{N}{2} p(1-p) \left(f\left(p - \frac{1}{N}\right) - f(p) \right)$$

$$+ N v_1 p \left(f\left(p - \frac{1}{N}\right) - f(p) \right) + N v_2 (1-p) \left(f\left(p + \frac{1}{N}\right) - f(p) \right).$$

$$(2.10)$$

To see this, note that the reproduction events in the Moran model take place at the points of a Poisson process with rate $\binom{N}{2}$ and at the time of such a transition there will only be a change in allele frequencies if the two individuals chosen to be involved in the reproduction event are of different allelic types. Thus, if the current proportion of a alleles in the population is p, then

$$p \mapsto p + \frac{1}{N} \quad \text{with probability } p(1-p),$$

$$p \mapsto p - \frac{1}{N} \quad \text{with probability } p(1-p)$$

and there is no change with probability $1 - 2p(1-p)$. The chance that we see a reproduction event in a time interval of length δt is

$$\binom{N}{2} \delta t + \mathscr{O}((\delta t)^2)$$

and the probability of seeing more than one transition is $\mathscr{O}((\delta t)^2)$. For mutation events, at total rate $N p v_1$, one of the Np type a individuals will mutate to type A, resulting in a reduction of p by $1/N$ and at total rate $N(1-p) v_2$ one of the $N(1-p)$ type A individuals will mutate to type a. Putting all this together gives that for $f : [0,1] \to \mathbb{R}$ and $p = i/N$ for some $i \in \{0, 1, \ldots, N\}$, $\mathscr{L}_N f(p)$ is given by (2.10).

To see what our population process will look like for large N we take f to be three times continuously differentiable, and use Taylor's Theorem to find an approximation for $\mathscr{L}_N f$. Thus

$$
\begin{aligned}
\mathscr{L}_N f(p) &= \binom{N}{2} p(1-p) \left(f(p) + \frac{1}{N} f'(p) + \frac{1}{2N^2} f''(p) + \mathscr{O}\left(\frac{1}{N^3}\right) - f(p) \right) \\
&\quad + \binom{N}{2} p(1-p) \left(f(p) - \frac{1}{N} f'(p) + \frac{1}{2N^2} f''(p) + \mathscr{O}\left(\frac{1}{N^3}\right) - f(p) \right) \\
&\quad + N p v_1 \left(f(p) - \frac{1}{N} f'(p) + \mathscr{O}\left(\frac{1}{N^2}\right) - f(p) \right) \\
&\quad + N(1-p) v_2 \left(f(p) + \frac{1}{N} f'(p) + \mathscr{O}\left(\frac{1}{N^2}\right) - f(p) \right) \\
&= \frac{1}{2} p(1-p) f''(p) + ((1-p) v_2 - p v_1) f'(p) + \mathscr{O}\left(\frac{1}{N}\right).
\end{aligned}
$$

We have proved the following.

Lemma 2.39. *As $N \to \infty$, the generator $\mathscr{L}_N$ of the process of allele frequencies under the neutral Moran model with mutation converges to $\mathscr{L}$, the generator of the Wright–Fisher diffusion with mutation, which is given by*

$$
\mathscr{L} f(p) = \frac{d}{dt} \mathbb{E}\left[f(p_t) | p_0 = p \right]\Big|_{t=0} = \frac{1}{2} p(1-p) f''(p) + (v_2 - (v_1 + v_2) p) f'(p).
$$
(2.11)

Remark 2.40. Notice, in particular, that if we set $v_1 = v_2 = 0$ we obtain

$$
\mathscr{L} f(p) = \frac{1}{2} p(1-p) f''(p),
$$

which is exactly the generator that we obtained in the large population limit from our Wright–Fisher model. It is not hard to extend the work that we did there to include mutations and recover the full generator (2.11).

What we have written down is the generator of a one-dimensional diffusion. We should like to be able to use the convergence of generators that we have verified to justify using the corresponding one-dimensional diffusion as an approximation to the process of allele frequencies under the Moran, the Wright–Fisher and the Cannings models (on suitable timescales). We defer the statement of a theorem that provides that justification until Sect. 3.2. Evidently we also need to know that there is a unique Markov process with generator (2.11) and that we can actually calculate quantities of interest for it. Happily both are true.

Chapter 3
One Dimensional Diffusions

3.1 Diffusions

In this chapter we are going to remind ourselves of some useful facts about one-dimensional diffusions. It is not an exhaustive study. Excellent references for this material are Karlin and Taylor (1981) and Knight (1981). We start in a fairly general setting.

Definition 3.1 (One-dimensional diffusion). A one-dimensional diffusion process $\{X_t\}_{t\geq 0}$ is a strong Markov process on $\mathbb{R}$ which traces out a continuous path as time evolves.

At any instant in time, X_t is a continuous random variable but also any realisation of $\{X_t\}_{t\geq 0}$ is a continuous function of time. Its range need not be the whole of $\mathbb{R}$ and indeed for the most part we'll be interested in diffusions on $(0,1)$ (possibly union either or both of the endpoints $\{0,1\}$). For the time being let us take the state space to be an interval (a,b) (possibly infinite), again possibly union one or both of the endpoints $\{a,b\}$.[1]

Remark 3.2. In our discussion of diffusions, we shall generally use x to denote a generic point in (a,b), but for consistency with our previous notation we shall use p for points in $(0,1)$ when discussing models for allele frequencies in genetics.

The generator of the diffusion takes the form

$$\mathscr{L}f(x) = \frac{1}{2}\sigma^2(x)\frac{d^2 f}{dx^2}(x) + \mu(x)\frac{df}{dx}(x). \tag{3.1}$$

Evidently for this to be defined f must be twice differentiable on (a,b). Depending on the behaviour of the diffusion close to the boundaries of its domain, f may also have to satisfy boundary conditions at a and b. We'll specify these precisely in Theorem 3.17, but for now assume that if we apply the generator to a function, f,

---

[1] We wish to include all *accessible* endpoints, defined in Definition 3.16, in the state space.

A. Etheridge, *Some Mathematical Models from Population Genetics*, Lecture Notes in Mathematics 2012, DOI 10.1007/978-3-642-16632-7_3,

then f is in the domain $\mathscr{D}(\mathscr{L})$ of the generator $\mathscr{L}$. To avoid pathologies, we make the following assumptions:

1. For any compact interval $I \subset (a,b)$, there exists $\varepsilon > 0$ such that $\sigma^2(x) > \varepsilon$ for all $x \in I$.
2. The coefficients $\mu(x)$ and $\sigma^2(x)$ are continuous functions of $x \in (a,b)$.

Note that (crucially for applications in genetics) we *do* allow $\sigma^2(x)$ to vanish at the boundary points $\{a,b\}$.

Let us write $\Delta_h X(t) = X_{t+h} - X_t$, then taking $f_1(x) = x$ in the generator (and using the Markov property) we see that

$$\mathscr{L} f_1(X_t) = \lim_{h \downarrow 0} \frac{1}{h} \mathbb{E}[\Delta_h X(t) | X_t] = \mu(X_t)$$

and so

$$\mathbb{E}[\Delta_h X(t) | X_t] = h\mu(X_t) + o(h) \qquad \text{as } h \downarrow 0. \tag{3.2}$$

Now observe that we can write $(X_{t+h} - X_t)^2 = X_{t+h}^2 - X_t^2 - 2X_t(X_{t+h} - X_t)$ and so, taking $f_2(x) = x^2$,

$$\mathscr{L} f_2(X_t) - 2X_t \mathscr{L} f_1(X_t) = \lim_{h \downarrow 0} \frac{1}{h} \mathbb{E}[(\Delta_h X(t))^2 | X_t] = \sigma^2(X_t),$$

which yields

$$\mathbb{E}[(\Delta_h X(t))^2 | X_t] = h\sigma^2(X_t) + o(h) \qquad \text{as } h \downarrow 0. \tag{3.3}$$

This motivates the following terminology.

Definition 3.3 (Infinitesimal drift and variance). The coefficients $\mu(x)$ and $\sigma^2(x)$ are called the (infinitesimal) *drift* and *variance* of the diffusion $\{X_t\}_{t \geq 0}$.

In fact, if a strong Markov process $\{X_t\}_{t \geq 0}$ is càdlàg (that is its paths are right continuous with left limits) and satisfies (3.2), (3.3) and the additional condition

$$\lim_{h \downarrow 0} \frac{1}{h} \mathbb{E}[|\Delta_h X(t)|^p | X_t = x] = 0 \qquad \text{for some } p > 2,$$

where the convergence is uniform in (x,t) on compact subsets of $(a,b) \times \mathbb{R}_+$, then $\{X_t\}_{t \geq 0}$ is necessarily a diffusion (see Karlin and Taylor (1981), Sect. 15.1, Lemma 1.1).

The canonical example of a one-dimensional diffusion is one-dimensional Brownian motion which has generator

$$\mathscr{L}_B f(x) = \frac{1}{2} \frac{d^2 f}{dx^2}(x).$$

It has transition density function

$$p(t,x,y) = \frac{1}{\sqrt{2\pi t}} \exp\left(-\frac{(x-y)^2}{2t}\right).$$

In other words, if $\{W_t\}_{t\geq 0}$ denotes Brownian motion then

$$\mathbb{P}[W_t \in A | X_0 = x] \equiv \mathbb{P}_x[W_t \in A] = \int_A \frac{1}{\sqrt{2\pi t}} \exp\left(-\frac{(x-y)^2}{2t}\right) dy$$

for any subset $A \subseteq \mathbb{R}$.

Brownian motion can be thought of as a building block from which other one-dimensional diffusions are constructed. One approach is to observe that the diffusion corresponding to the generator $\mathscr{L}$ of (3.1) can be expressed as the solution of a stochastic differential equation driven by Brownian motion (with appropriate boundary conditions)

$$dX_t = \mu(X_t)dt + \sigma(X_t)dW_t. \tag{3.4}$$

Remark 3.4 (Mathematical drift versus genetic drift). We have already encountered the Wright–Fisher diffusion several times, corresponding to the solution of the stochastic differential equation

$$dp_t = (v_2(1-p_t) - v_1 p_t)dt + \sqrt{p_t(1-p_t)}dW_t.$$

It is an unfortunate accident of history that the standard terminology for the stochastic term (driven by Brownian motion) is *genetic drift*, whereas to a mathematician it is the *deterministic* mutation term that corresponds to drift.

We can see from (3.4) and Itô's Lemma that

$$f(X_t) - \int_0^t \mathscr{L}f(X_s)ds \tag{3.5}$$

is a martingale for all $f \in \mathscr{D}(\mathscr{L})$. Stroock and Varadhan (1979) use this martingale property as a way of characterising the Markov process associated with a given generator. Solving the corresponding martingale problem requires all the expressions in (3.5) to be martingales.

Definition 3.5. We shall say that the *martingale problem* for μ, σ is well-posed if for each $x \in (a,b)$ there is a unique probability measure $\mathbb{P}_x$ on the continuous functions from $[0,1]$ to $\mathbb{R}$ (with the σ-field generated by the coordinate maps) such that $\mathbb{P}[X_0 = x] = 1$ and the quantities in (3.5) are martingales.

We refer to Stroock and Varadhan (1979) for a thorough introduction to martingale problems. In particular, the martingale problem for a diffusion process on $\mathbb{R}$ is certainly well-posed if μ and σ^2 are bounded measurable functions with σ^2 uniformly strictly positive. (This last condition is violated by the Wright–Fisher diffusion, but it turns out that the martingale problem is nonetheless well-posed, see Sect. 3.2 for references.)

Our approach to constructing one-dimensional diffusions from Brownian motion will not be via stochastic differential equations or martingale problems, but rather through the theory of speed and scale. Before introducing that, let's fill a gap that we left at the end of Chap. 2.

3.2 Convergence to Diffusions

In the case where the martingale problem of Definition 3.5 is well-posed, Stroock and Varadhan (1979) provide elementary criteria for convergence of discrete or continuous time Markov chains to a diffusion, which we record in Theorem 3.6 below. We follow Sect. 8.7 of Durrett (1996) which treats discrete and continuous time together. We need some notation. Suppose that we have a series of discrete time Markov chains, $\{Y_{nh}^h\}_{n \in \mathbb{N}}$ say, indexed by h and taking values in $S_h \subseteq \mathbb{R}$, with the chain with index h jumping at time intervals of length h. Write

$$\mathbb{P}[Y_{(n+1)h}^h \in A | Y_{nh}^h = x] = \Pi_h(x, A), \qquad \text{for } x \in S_h, A \subset \mathbb{R}.$$

(When we write $A \subseteq \mathbb{R}$ we implicitly assume that A is a Borel subset of $\mathbb{R}$.) We define $X_t^h = Y_{h[t/h]}^h$, where $[u]$ denotes the integer part of $u \in \mathbb{R}$. In other words we extend Y^h to all times $t \geq 0$ by setting it to be constant on time intervals $[nh, (n+1)h)$.

Now suppose that we have continuous time chains $\{X_t^h\}_{t \geq 0}$ taking values in $S_h \subseteq \mathbb{R}$. In place of the sequence of transition probabilities Π_h for the discrete time chain, we have a sequence of transition rates:

$$\left. \frac{d}{dt} \mathbb{P}[X_t^h \in A | X_0^h = x] \right|_{t=0} = Q_h(x, A), \qquad \text{for } x \in S_h, A \subset \mathbb{R}, x \notin A.$$

We assume that for any compact set K,

$$\sup_{x \in K} Q_h(x, \mathbb{R}) < \infty. \tag{3.6}$$

Let us write

$$K_h(x, dy) = \begin{cases} \frac{1}{h} \Pi_h(x, dy) & \text{in discrete time} \\ Q_h(x, dy) & \text{in continuous time} \end{cases}$$

and define

$$(\sigma^2)^h(x) = \int_{|y-x| \leq 1} (y-x)^2 K_h(x, dy),$$

$$\mu^h(x) = \int_{|y-x| \le 1} (y-x) K_h(x, dy),$$

$$\Delta_\varepsilon^h(x) = K_h(x, B(x, \varepsilon)^c),$$

where $B(x, \varepsilon) = (x - \varepsilon, x + \varepsilon)$.

Theorem 3.6. *Suppose that μ and σ are continuous coefficients for which the martingale problem for $\mathscr{L}$ of (3.1) is well-posed. In continuous time we assume (3.6). Suppose further that for each $R < \infty$ and $\varepsilon > 0$*

1.

$$\lim_{h \downarrow 0} \sup_{x \in S_h, |x| \le R} |\mu^h(x) - \mu(x)| = 0,$$

2.

$$\lim_{h \downarrow 0} \sup_{x \in S_h, |x| \le R} |(\sigma^2)^h(x) - \sigma^2(x)| = 0,$$

3.

$$\lim_{h \downarrow 0} \sup_{x \in S_h, |x| \le R} \Delta_\varepsilon^h(x) = 0.$$

If $X_0^h = x_h \to x$ then we have $\{X_t^h\}_{t \ge 0} \Rightarrow \{X_t\}_{t \ge 0}$, the solution of the martingale problem with $X_0 = x$. (Here $\Rightarrow$ denotes convergence in the sense of finite-dimensional distributions.)

This is a special case of Theorem 8.7.1 of Durrett (1996) which in turn is based upon Chap. 11 of Stroock and Varadhan (1979). The first two conditions of the Theorem ensure that infinitesimal drift and variance of the sequence of Markov chains converge (uniformly on compact sets) to the right thing, while the third rules out jumps in the limit.

Of course it remains to check that the martingale problem *is* well-posed for our Wright–Fisher diffusion. That result is really due to Feller (1951) (although he didn't use this language). It can be found in Ethier and Kurtz (1986) who consider convergence of a Wright–Fisher model (with possibly more than two alleles) to the Wright–Fisher diffusion in their Chap. 10. They invoke much more powerful weak convergence results that are beyond our scope here.

Remark 3.7. This sort of convergence is enough to justify using our limiting Wright–Fisher diffusion to approximate things like time to fixation and fixation probabilities. However, if we are really interested in the genealogies of populations, then we need more. For our Moran models, the Donnelly–Kurtz lookdown construction gave us a much stronger result, namely the joint convergence of the forwards in time model for the evolution of the population and the (backwards in time) genealogical trees relating individuals in that population. In general we must be careful. It is possible to arrive at the same diffusion for allele frequencies from many different individual based models for our population, and it is *not* always the case that the genealogies converge to the same limit (see Taylor (2009) for some examples).

3.3   Speed and Scale

A nice feature of one dimensional diffusions is that many quantities can be calculated explicitly. This is because (except at certain singular points which will only ever be at a or b under our conditions) all one-dimensional diffusions can be transformed into Brownian motion first by a change of space variable (through the so-called scale function) and then a timechange (through what is known as the speed measure).

To see how this works, we first investigate what happens to a diffusion when we change the timescale. Suppose that a diffusion $\{Z_t\}_{t\geq 0}$ has generator $\mathscr{L}_Z$, with infinitesimal drift $\mu_Z(x)$ and infinitesimal variance $\sigma_Z^2(x)$. We define a new process $\{Y_t\}_{t\geq 0}$ by $Y_t = Z_{\tau(t)}$ where

$$\tau(t) = \int_0^t \beta(Y_s)ds,$$

for some function $\beta(x)$ which we assume to be bounded, continuous and strictly positive. So if $Y_0 = Z_0$, then the increment of Y_t over an infinitesimal time interval $(0,dt)$ is that of Z_t over the interval $(0,d\tau(t)) = (0,\beta(Y_0)dt)$. In our previous notation,

$$\mathbb{E}[\Delta_h Y(0)|Y_0 = y] = \beta(Y_0)h\mu_Z(Z_0) + o(h) = \beta(y)\mu_Z(y)h + o(h),$$

and

$$\mathbb{E}[(\Delta_h Y(0))^2|Y_0 = y] = \beta(Y_0)h\sigma_Z^2(Z_0) + o(h) = \beta(y)\sigma_Z^2(y)h + o(h).$$

In other words,

$$\mathscr{L}_Y f(x) = \beta(x)\mathscr{L}_Z f(x).$$

In the simplest example, β is a constant and we are simply changing our time units in a spatially homogeneous way. In general, the rate of our 'clock' depends upon where we are in space. We are now in a position to understand speed and scale. Let $\{X_t\}_{t\geq 0}$ be governed by the generator (3.1). Suppose now that $S(x)$ is a strictly increasing function on (a,b) and consider the new process $Z_t = S(X_t)$. Then the generator $\mathscr{L}_Z$ of Z can be calculated as

$$
\begin{aligned}
\mathscr{L}_Z f(x) &= \frac{d}{dt}\mathbb{E}[f(Z_t)|Z_0 = x]\Big|_{t=0} \\
&= \frac{d}{dt}\mathbb{E}[f(S(X_t))|S(X_0) = x]\Big|_{t=0} \\
&= \mathscr{L}_X(f \circ S)(S^{-1}(x)) \\
&= \frac{1}{2}\sigma^2(S^{-1}(x))\frac{d^2}{dx^2}(f \circ S)(S^{-1}(x)) + \mu(S^{-1}(x))\frac{d}{dx}(f \circ S)(S^{-1}(x))
\end{aligned}
$$

$$= \frac{1}{2}\sigma^2(S^{-1}(x))\left\{(S'(S^{-1}(x)))^2\frac{d^2f}{dx^2}(x) + S''(S^{-1}(x))\frac{df}{dx}(x)\right\}$$

$$+ \mu(S^{-1}(x))S'(S^{-1}(x))\frac{df}{dx}(x)$$

$$= \frac{1}{2}\sigma^2(S^{-1}(x))S'(S^{-1}(x))^2\frac{d^2f}{dx^2}(x) + \mathscr{L}S(S^{-1}(x))\frac{df}{dx}(x). \tag{3.7}$$

Now if we can find a strictly increasing function S that satisfies $\mathscr{L}S \equiv 0$, then the drift term (in the mathematical sense) in (3.7) will vanish and so Z_t will just be a time change of Brownian motion on the interval $(S(a), S(b))$. Such an S is provided by the scale function of the diffusion.

Definition 3.8 (Scale function). For a diffusion X_t on (a, b) with drift μ and variance σ^2, the *scale function* is defined by

$$S(x) = \int_{x_0}^x \exp\left(-\int_\eta^y \frac{2\mu(z)}{\sigma^2(z)}dz\right)dy,$$

where x_0, η are points fixed (arbitrarily) in (a, b).

Definition 3.9 (Natural scale). We shall say that a diffusion is in *natural scale* if $S(x)$ can be taken to be linear.

The scale change $X_t \mapsto S(X_t)$ resulted in a timechanged Brownian motion on $(S(a), S(b))$. The change of time required to transform this into standard Brownian motion is dictated by the speed measure.

Definition 3.10 (Speed measure). The function $m(\xi) = \frac{1}{\sigma^2(\xi)S'(\xi)}$ is the *density of the speed measure* or just the *speed density* of the process X_t. We write

$$M(x) = \int_{x_0}^x m(\xi)d\xi.$$

Remark 3.11. The function m plays the rôle of β before. Naively, looking at (3.7), we might expect to timechange via $\beta(\xi) = 1/(\sigma^2(\xi)S'(\xi)^2)$. However, notice that

$$\int_{x_0}^x m(\xi)d\xi = \int_{S(x_0)}^{S(x)} m(S^{-1}(y))\frac{1}{S'(S^{-1}(y))}dy = \int_{S(x_0)}^{S(x)} \frac{1}{\sigma^2(S^{-1}(y))(S'(S^{-1}(y)))^2}dy.$$

The additional $S'(y)$ in the generator (3.7) has been absorbed in the change of coordinates since our time change is applied to $S(X_t)$ on $(S(a), S(b))$, not to X_t itself.

In summary, we have the following.

Lemma 3.12. *Denoting the scale function and the speed measure by S and M respectively we have*

$$\mathscr{L}f = \frac{1}{2}\frac{1}{dM/dS}\frac{d^2f}{dS^2} = \frac{1}{2}\frac{d}{dM}\left(\frac{df}{dS}\right).$$

Proof.

$$\frac{1}{2}\frac{d}{dM}\left(\frac{df}{dS}\right) = \frac{1}{2}\frac{1}{dM/dx}\frac{d}{dx}\left(\frac{1}{dS/dx}\frac{df}{dx}\right)$$

$$= \frac{1}{2}\sigma^2(x)S'(x)\frac{d}{dx}\left(\frac{1}{S'(x)}\frac{df}{dx}\right)$$

$$= \frac{1}{2}\sigma^2(x)\frac{d^2f}{dx^2} - \frac{1}{2}\sigma^2(x)S'(x)\frac{S''(x)}{(S'(x))^2}\frac{df}{dx}$$

$$= \frac{1}{2}\sigma^2(x)\frac{d^2f}{dx^2} + \mu(x)\frac{df}{dx}$$

(since S solves $\mathscr{L}S = 0$) as required.   □

3.4 Hitting Probabilities and Feller's Boundary Classification

Before going further, let's see how we might apply this. Suppose that a diffusion process on $(0,1)$ represents the frequency of an allele, a say, in a population and that zero and one are traps for the process. One question that we should like to answer is "What is the probability that the a-allele is eventually lost from the population?" In other words, what is the probability that the diffusion hits zero before one? To prove a general result we need first to be able to answer this question for Brownian motion.

Lemma 3.13. *Let $\{W_t\}_{t\geq 0}$ be standard Brownian motion on the line. For each $y \in \mathbb{R}$, let T_y denote the random time at which it hits y for the first time. Then for $a < x < b$,*

$$\mathbb{P}[T_a < T_b | W_0 = x] = \frac{b-x}{b-a}.$$

Sketch of Proof. Let $u(x) = \mathbb{P}[T_a < T_b | W_0 = x]$ and assume that $\mathbb{P}[T_a \wedge T_b < h | W_0 = x] = o(h)$ as $h \to 0$. If we suppose that u is sufficiently smooth, then, using the Markov property,

$$u(x) = \mathbb{E}[u(W_h) | W_0 = x] + o(h)$$

$$= \mathbb{E}\left[u(x) + (W_h - x)u'(x) + \frac{1}{2}(W_h - x)^2 u''(x)\right] + o(h)$$

$$= u(x) + \frac{1}{2}hu''(x) + o(h).$$

Subtracting $u(x)$ from each side, dividing by h and letting h tend to zero, we obtain $u''(x) = 0$. We also have the boundary conditions $u(a) = 1$ and $u(b) = 0$. This is easily solved to give

$$u(x) = \frac{b-x}{b-a}$$

as required.   □

Of course this reflects the corresponding result for simple random walk that we used in the proof of Lemma 2.31. In general we can reduce the corresponding question for $\{X_t\}_{t \geq 0}$ to solution of the equation $\mathscr{L}u(x) = 0$ with $u(a) = 1$ and $u(b) = 0$, but in fact we have already done all the work we need. We have the following result.

Lemma 3.14 (Hitting probabilities). *Let $\{X_t\}_{t \geq 0}$ be a one-dimensional diffusion on (a, b) with infinitesimal drift $\mu(x)$ and variance $\sigma^2(x)$ satisfying the conditions above. If $a < a_0 < x < b_0 < b$ then writing T_y for the first time at which $X_t = y$,*

$$\mathbb{P}[T_{a_0} < T_{b_0} | X_0 = x] = \frac{S(b_0) - S(x)}{S(b_0) - S(a_0)}, \tag{3.8}$$

where S is the scale function for the diffusion.

Remark 3.15. Our definition of the scale function, S, depended upon arbitrary choices of η and x_0, but η cancels in the ratio and x_0 in the difference, so that the expression on the right hand side of (3.8) is well-defined.

Proof. Evidently it is enough to consider the corresponding hitting probabilities for the process $Z_t = S(X_t)$, where S is the scale function. The process $\{Z_t\}_{t \geq 0}$ is a time changed Brownian motion, but since we only care about *where* not *when* the process exits the interval $(S(a_0), S(b_0))$, then we need only determine the hitting probabilities for Brownian motion and the result follows immediately from Lemma 3.13.   □

Before continuing to calculate quantities of interest, we fill in a gap left earlier when we failed to completely specify the domain of the generators of our one-dimensional diffusions. Whether or not functions in the domain must satisfy boundary conditions at a and b is determined by the nature of those boundaries from the perspective of the diffusion. More precisely, we have the following classification.

Definition 3.16 (Feller's boundary classification). For a one-dimensional diffusion on the interval with endpoints a, b (with $a < b$), define

$$u(x) = \int_{x_0}^{x} M dS, \qquad v(x) = \int_{x_0}^{x} S dM,$$

where S is the scale function of Definition 3.8 and M the speed measure of Definition 3.10. The boundary b is said to be

$$\text{regular} \quad \text{if } u(b) < \infty \text{ and } v(b) < \infty$$
$$\text{exit} \quad\quad \text{if } u(b) < \infty \text{ and } v(b) = \infty$$
$$\text{entrance if } u(b) = \infty \text{ and } v(b) < \infty$$
$$\text{natural} \quad \text{if } u(b) = \infty \text{ and } v(b) = \infty$$

with symmetric definitions at a.

Regular and exit boundaries are said to be *accessible* while entrance and natural boundaries are called *inaccessible*.

Theorem 3.17. *If neither a nor b is regular, the domain of the generator (3.1) is the continuous functions f on $[a,b]$ which are twice continuously differentiable on the interior and for which*

1. *if a and b are inaccessible there are no further conditions,*
2. *if b (resp. a) is an exit boundary, then*

$$\lim_{x \to b} \mathscr{L} f(x) = 0$$

$$(\textit{resp. } \lim_{x \to a} \mathscr{L} f(x) = 0).$$

If b (resp. a) is a regular boundary, then for each fixed $q \in [0,1]$ there is a Feller semigroup corresponding to the generator (3.1) with domain as above plus the additional condition

$$q \lim_{x \to b} \mathscr{L} f(x) = (1-q) \lim_{x \to b} \frac{1}{S'(x)} f'(x) \tag{3.9}$$

$$\left(\textit{resp. } q \lim_{x \to a} \mathscr{L} f(x) = -(1-q) \lim_{x \to a} \frac{1}{S'(x)} f'(x) \right).$$

For a more careful discussion see Ethier and Kurtz (1986), Chap. 8.

Remark 3.18. For each fixed $q \in [0,1]$, condition (3.9) is enough to specify the boundary behaviour of the diffusion at a regular boundary uniquely. It is easy to check that the Wright–Fisher diffusion with mutation with generator (2.11) has a regular boundary at 0 (resp. 1) if $v_2 \in (0, \frac{1}{2})$ (resp. if $v_1 \in (0, \frac{1}{2})$), but the condition (3.9) is in fact *the same* for all $q > 0$.

3.5   Green's Functions

Lemma 3.14 tells us the probability that we exit (a,b) for the first time through a, but can we glean some information about how long we must wait for $\{X_t\}_{t \geq 0}$ to exit the interval (a,b) (either through a or b) or, more generally, writing T^* for the first exit time of (a,b), can we say anything about $\mathbb{E}[\int_0^{T^*} g(X_s)ds | X_0 = x]$? (Putting $g = 1$ gives the mean exit time.) Let us write

$$w(x) = \mathbb{E}\left[\int_0^{T^*} g(X_s)ds|X_0 = x\right]$$

and we'll derive the differential equation satisfied by w.

Suppose for simplicity that g is Lipschitz continuous on (a,b) with Lipschitz constant K. First note that $w(a) = w(b) = 0$. Now consider a small interval of time of length h. We're going to split the integral into the contribution up to time h and after time h. Because $\{X_t\}_{t \geq 0}$ is a Markov process,

$$\mathbb{E}\left[\int_h^{T^*} g(X_s)ds|X_h = z\right] = \mathbb{E}\left[\int_0^{T^*} g(X_s)ds|X_0 = z\right] = w(z)$$

and so for $a < x < b$

$$w(x) \approx \mathbb{E}\left[\int_0^h g(X_s)ds|X_0 = x\right] + \mathbb{E}\left[w(X_h)|X_0 = x\right]. \tag{3.10}$$

The '$\approx$' here is because we have ignored the possibility that $h > T^*$. Since g is Lipschitz continuous, we have the approximation

$$\left|\mathbb{E}\left[\int_0^h g(X_s)ds|X_0 = x\right] - hg(x)\right| = \mathbb{E}\left[\left|\int_0^h g(X_s)ds - hg(x)\right|\Big|X_0 = x\right]$$

$$\leq \mathbb{E}\left[\int_0^h K|X_s - x|ds|X_0 = x\right] \leq K\int_0^h \sqrt{\mathbb{E}[|X_s - x|^2|X_0 = x]} = \mathcal{O}(h^{3/2}).$$

Now substitute this estimate in (3.10), subtract $w(x)$ from both sides, divide by h and let $h \downarrow 0$ to obtain

$$\mu(x)w'(x) + \frac{1}{2}\sigma^2(x)w''(x) = -g(x), \quad w(a) = 0 = w(b). \tag{3.11}$$

Let us now turn to solving this equation. Using Lemma 3.12 with $w = f$,

$$\mathcal{L}w(x) = \frac{1}{2}\frac{1}{m(x)}\frac{d}{dx}\left(\frac{1}{S'(x)}w'(x)\right)$$

and so we have

$$\frac{d}{dx}\left(\frac{1}{S'(x)}w'(x)\right) = -2g(x)m(x),$$

whence

$$\frac{1}{S'(x)}w'(x) = -2\int_a^x g(\xi)m(\xi)d\xi + \beta$$

where β is a constant. Multiplying by $S'(x)$ and integrating gives

$$w(x) = -2\int_a^x S'(\xi)\int_a^\xi g(\eta)m(\eta)d\eta d\xi + \beta(S(x) - S(a)) + \alpha$$

for constants α, β. Since $w(a) = 0$, we immediately have that $\alpha = 0$. Reversing the order of integration,

$$w(x) = -2 \int_a^x \int_\eta^x S'(\xi)d\xi g(\eta)m(\eta)d\eta + \beta(S(x) - S(a))$$

$$= -2 \int_a^x (S(x) - S(\eta))g(\eta)m(\eta)d\eta + \beta(S(x) - S(a))$$

and $w(b) = 0$ now gives

$$\beta = \frac{2}{S(b) - S(a)} \int_a^b (S(b) - S(\eta))g(\eta)m(\eta)d\eta.$$

Finally then

$$w(x) = \frac{2}{S(b) - S(a)} \left\{ (S(x) - S(a)) \int_a^b (S(b) - S(\eta))g(\eta)m(\eta)d\eta \right.$$

$$\left. -(S(b) - S(a)) \int_a^x (S(x) - S(\eta))g(\eta)m(\eta)d\eta \right\}$$

$$= \frac{2}{S(b) - S(a)} \left\{ (S(x) - S(a)) \int_x^b (S(b) - S(\eta))g(\eta)m(\eta)d\eta \right.$$

$$\left. +(S(b) - S(x)) \int_a^x (S(\eta) - S(a))g(\eta)m(\eta)d\eta \right\}$$

where the last line is obtained by splitting the first integral into $\int_a^b = \int_x^b + \int_a^x$.

Theorem 3.19. *For a continuous function g,*

$$\mathbb{E} \left[\int_0^{T*} g(X_s)ds | X_0 = x \right] = \int_a^b G(x, \xi)g(\xi)d\xi,$$

where for $a < x < b$ we have

$$G(x, \xi) = \begin{cases} 2\dfrac{(S(x) - S(a))}{(S(b) - S(a))}(S(b) - S(\xi))m(\xi), \textit{for } x < \xi < b \\[4mm] 2\dfrac{(S(b) - S(x))}{(S(b) - S(a))}(S(\xi) - S(a))m(\xi), \textit{for } a < \xi < x, \end{cases}$$

with S the scale function given in Definition 3.8 and $m(\xi) = \frac{1}{\sigma^2(\xi)S'(\xi)}$, the density of the speed measure.

Definition 3.20. The function $G(x, \xi)$ is called the *Green's function* of the process $\{X_t\}_{t \geq 0}$.

By taking g to approximate $\mathbf{1}_{(x_1, x_2)}$ we see that $\int_{x_1}^{x_2} G(x, \xi) d\xi$ is the mean time spent by the process in (x_1, x_2) before exiting (a, b) if initially $X_0 = x$. Sometimes, the Green's function is called the *sojourn density*.

Example 3.21. Consider the Wright–Fisher diffusion with generator

$$\mathscr{L}f(p) = \frac{1}{2}p(1-p)f''(p).$$

Notice that since it has no drift term ($\mu = 0$) it is already in natural scale, $S(p) = p$ (up to an arbitrary additive constant). What about $\mathbb{E}[T^*]$?

Using Theorem 3.19 with $g = 1$ we have

$$\mathbb{E}_p[T^*] = \mathbb{E}\left[\int_0^{T^*} 1 ds \middle| X_0 = p\right] = \int_0^1 G(p, \xi) d\xi$$

$$= 2\int_p^1 p(1-\xi)\frac{1}{\xi(1-\xi)}d\xi + 2\int_0^p (1-p)\xi\frac{1}{\xi(1-\xi)}d\xi$$

$$= 2p\int_p^1 \frac{1}{\xi}d\xi + 2(1-p)\int_0^p \frac{1}{1-\xi}d\xi$$

$$= -2\{p\log p + (1-p)\log(1-p)\}.$$

$\square$

This suggests that in our Moran model, at least if the population is large, if the current proportion of a-alleles is p, the time until either the a-allele or the A-allele is fixed in the population should have mean approximately

$$-2\{p\log p + (1-p)\log(1-p)\}. \tag{3.12}$$

In fact by conditioning on whether the proportion of a-alleles increases or decreases at the first reproduction event, one obtains a recurrence relation for the *number of jumps* until the Moran process first hits either zero or one. This recurrence relation can be solved explicitly and since jumps occur at independent exponentially distributed times with mean $1/\binom{N}{2}$, it is easy to verify that (3.12) is indeed a good approximation. For the Wright–Fisher model, in its original timescale, there is no explicit expression for the expected time to fixation, $t(p)$. However, since changes in p over a single generation are typically small, one can expand $t(p)$ in a Taylor series, in just the same way as we did to derive equation (2.1), and thus verify that for a large population,

$$p(1-p)t''(p) = -2N, \quad t(0) = 0 = t(1).$$

This is readily solved to give

$$t(p) = -2N\{p\log p + (1-p)\log(1-p)\},$$

as predicted by our diffusion approximation. (Recall that our Moran model is already in the diffusive timescale, whereas the Wright–Fisher model is not, accounting for the extra factor of N.)

3.6 Stationary Distributions and Reversibility

Before moving on to models in which a gene is allowed to have more than two alleles, we consider one last quantity for our one-dimensional diffusions. First a general definition.

Definition 3.22 (Stationary distribution). Let $\{X_t\}_{t\geq 0}$ be a Markov process on the space E. A *stationary distribution* for $\{X_t\}_{t\geq 0}$ is a probability distribution ψ on E such that if X_0 has distribution ψ, then X_t has distribution ψ for all $t \geq 0$.

In particular this definition tells us that if ψ is a stationary distribution for $\{X_t\}_{t\geq 0}$, then

$$\frac{d}{dt}\mathbb{E}[f(X_t)|X_0 \sim \psi] = 0,$$

where we have used $X_0 \sim \psi$ to indicate that X_0 is distributed according to ψ. In other words

$$\frac{d}{dt}\int_E \mathbb{E}[f(X_t)|X_0 = x]\,\psi(dx) = 0.$$

Evaluating the time derivative at $t = 0$ gives

$$\int_E \mathscr{L}f(x)\psi(dx) = 0.$$

Sometimes this allows us to find an explicit expression for $\psi(dx)$. Let $\{X_t\}_{t\geq 0}$ be a one-dimensional diffusion on (a,b) with generator given by (3.1). We're going to suppose that there is a stationary distribution which is absolutely continuous with respect to Lebesgue measure. Let us abuse notation a little by using $\psi(x)$ to denote the density of $\psi(dx)$ on (a,b). Then, integrating by parts, we have that for all $f \in \mathscr{D}(\mathscr{L})$,

$$0 = \int_a^b \left\{\frac{1}{2}\sigma^2(x)\frac{d^2f}{dx^2}(x) + \mu(x)\frac{df}{dx}(x)\right\}\psi(x)dx$$

$$= \int_a^b f(x)\left\{\frac{1}{2}\frac{d^2}{dx^2}\left(\sigma^2(x)\psi(x)\right) - \frac{d}{dx}\left(\mu(x)\psi(x)\right)\right\}dx + \text{boundary terms}.$$

This equation must hold for all f in the domain of $\mathscr{L}$ and so, in particular, choosing f and f' to vanish on the boundary,

$$\frac{1}{2}\frac{d^2}{dx^2}\left(\sigma^2(x)\psi(x)\right) - \frac{d}{dx}\left(\mu(x)\psi(x)\right) = 0 \quad \text{for } x \in (a,b). \tag{3.13}$$

Integrating once gives

$$\frac{1}{2}\frac{d}{dx}\left(\sigma^2(x)\psi(x)\right) - \mu(x)\psi(x) = C_1,$$

for some constant C_1 and then using $S'(x)$ as an integrating factor we obtain

$$\frac{d}{dy}\left(S'(y)\sigma^2(y)\psi(y)\right) = C_1 S'(y),$$

from which

$$\psi(x) = C_1\frac{S(x)}{S'(x)\sigma^2(x)} + C_2\frac{1}{S'(x)\sigma^2(x)} = m(x)\left[C_1 S(x) + C_2\right].$$

If we can arrange constants so that $\psi \geq 0$ and

$$\int_a^b \psi(\xi)d\xi = 1$$

then the stationary distribution exists and has density ψ. In particular, if $\int_a^b m(y)\,dy < \infty$, then taking $C_1 = 0$,

$$\psi(x) = \frac{m(x)}{\int_a^b m(y)dy} \tag{3.14}$$

is the density of a stationary distribution for the diffusion.

We know from the theory of Markov chains that uniqueness of the stationary measure of a chain requires irreducibility. The corresponding condition here is regularity.

Definition 3.23. For a one dimensional diffusion process on the interval I, let us write

$$H_y = \inf\{t > 0 : X_t = y\}.$$

The diffusion is said to be *regular* if for all $x \in I^0$ (the interior of I) and all $y \in I$ (including finite endpoints) $\mathbb{P}_x[H_y < \infty] > 0$.

Theorem 3.24 (Watanabe and Motoo 1958). *A regular diffusion in natural scale with no absorbing boundary points has a stationary distribution if and only if the speed measure is finite and then it is given by (3.14).*

Under these conditions there is also an ergodic theorem.

Example 3.25. Recall the generator of the Wright–Fisher diffusion with mutation,

$$\mathscr{L}f(p) = \frac{1}{2}p(1-p)\frac{d^2f}{dp^2} + \left(v_2(1-p) - v_1 p\right)\frac{df}{dp}.$$

What is the stationary distribution?

For this diffusion

$$S'(p) = \exp\left(-\int_{p_0}^p \frac{2\mu(z)}{\sigma^2(z)}dz\right)$$

$$= \exp\left(-\int_{p_0}^p \frac{2v_2(1-z) - 2v_1 z}{z(1-z)}dz\right)$$

$$= C\exp\left(-2v_2\log p - 2v_1\log(1-p)\right)$$

$$= Cp^{-2v_2}(1-p)^{-2v_1},$$

where the value of the constant C depends on p_0. In this case we have

$$m(p) = \frac{1}{\sigma^2(p)S'(p)} = Cp^{2v_2-1}(1-p)^{2v_1-1}.$$

Now

$$\int_0^1 m(p)dp = \int_0^1 Cp^{2v_2-1}(1-p)^{2v_1-1}dp = C\frac{\Gamma(2v_1)\Gamma(2v_2)}{\Gamma(2(v_1+v_2))}$$

(where Γ is Euler's Gamma function) and so the stationary distribution is just

$$\psi(p) = \frac{\Gamma(2(v_1+v_2))}{\Gamma(2v_1)\Gamma(2v_2)}p^{2v_2-1}(1-p)^{2v_1-1}. \tag{3.15}$$

Ethier and Kurtz (1986), Chap. 10, Lemma 2.1 gives a direct proof of uniqueness of this stationary distribution. $\qquad\square$

The stationary distribution gives us some understanding of the longterm balance between the competing forces of mutation (which maintains genetic diversity) and genetic drift (which removes variation from the population). Figure 3.1 shows the density of the stationary distribution of the Wright–Fisher diffusion with mutation for a variety of parameters. When $2v_1$ and $2v_2$ are both bigger than 1, the stationary distribution is peaked around its mean, but when they are both less than one it has singularities at $\{0, 1\}$. Of course, if there is no mutation, then the process eventually becomes trapped in 0 and 1.

One can also calculate simple summary statistics.

Definition 3.26. The *gene diversity* or *heterozygosity*, H, is the probability that two randomly chosen genes are of different allelic types.

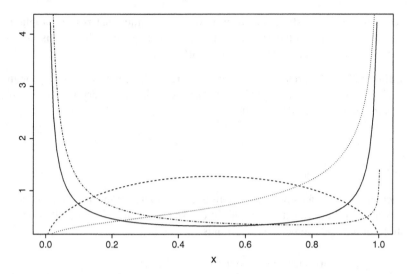

Fig. 3.1 Stationary distribution of the Wright–Fisher diffusion. The graphs plot the density ψ, given by (3.15) for: $2v_1 = 2v_2 = 0.2$ (*solid line*), $2v_1 = 2v_2 = 1.5$ (*dashed line*), $2v_1 = 0.5, 2v_2 = 1.3$ (*dotted line*) and $2v_1 = 0.7, 2v_2 = 0.2$ (alternating *dashes* and *dots*)

If the allele frequency, P say, is at stationarity under the Wright–Fisher diffusion with mutation, then

$$H = \mathbb{E}[2P(1-P)] = \int_0^1 2p(1-p)\psi(p)\,dp$$

$$= 2\int_0^1 p^{2v_2}(1-p)^{2v_1}dp \frac{\Gamma(2(v_1+v_2))}{\Gamma(2v_1)\Gamma(2v_2)}$$

$$= 2\frac{\Gamma(2v_1+1)\Gamma(2v_2+1)}{\Gamma(2v_1+2v_2+2)} \frac{\Gamma(2(v_1+v_2))}{\Gamma(2v_1)\Gamma(2v_2)}$$

$$= \frac{2 \cdot 2v_1 \cdot 2v_2}{(2v_1+2v_2)(2v_1+2v_2+1)}$$

$$= \frac{4v_1 v_2}{(v_1+v_2)(2(v_1+v_2)+1)}.$$

Now in 'real' units, $v_1 = N\mu_1$ and $v_2 = N\mu_2$ (or $2N_e\mu_1$ and $2N_e\mu_2$ for a diploid population) and so

$$H = \frac{2\mu_1\mu_2}{(\mu_1+\mu_2)(\mu_1+\mu_2+\frac{1}{2N})}.$$

Notice in particular that gene diversity increases with population size. For a larger population the force of genetic drift is weaker.

Finally let us demonstrate one very powerful technique that is often applied in settings where the speed measure is a stationary distribution. The idea is familiar from the study of discrete time and space Markov chains.

Definition 3.27. A discrete time and space Markov chain with transition probabilities $p(i, j)$ is said to be *reversible* with respect to the stationary measure π if it satisfies the *detailed balance* equation:

$$\pi(i)p(i, j) = \pi(j)p(j, i)$$

for all i and j in the state space.

For such chains we can say things about events backwards in time by considering the forwards in time transition probabilities. The analogue of the detailed balance equation for a one-dimensional diffusion is

$$\psi(x)p(t, x, y) = \psi(y)p(t, y, x) \quad \text{for all } x, y, t.$$

Now multiplying by arbitrary functions $f(x)$ and $g(y)$ in the domain of the generator of the diffusion we obtain

$$\int \psi(x)f(x) \left(\int p(t, x, y)g(y)dy \right) dx = \int \psi(y)g(y) \left(\int p(t, y, x)f(x)dx \right) dy.$$

Now observe that the inner integrals are

$$\mathbb{E}[g(X_t)|X_0 = x] \quad \text{and} \quad \mathbb{E}[f(X_t)|X_0 = y]$$

and differentiate with respect to t at $t = 0$ to obtain

$$\int f(x)\mathscr{L}g(x)\psi(x)dx = \int \mathscr{L}f(y)g(y)\psi(y)dy. \tag{3.16}$$

Definition 3.28. If the identity (3.16) is satisfied for all f and g, then ψ is called a *reversible stationary distribution* and we say that the diffusion is *reversible* with respect to ψ.

Now suppose that the stationary distribution of the diffusion is given by $\psi(x) = m(x)/\int m(y)dy$. Then choosing f and g to vanish at the boundary of the domain to force the boundary terms to vanish when we integrate by parts (twice), we obtain

$$\int_a^b f(x)\mathscr{L}g(x)m(x)dx = \frac{1}{2} \int_a^b f(x)\frac{1}{m(x)}\frac{d}{dx}\left(\frac{1}{S'(x)}\frac{dg}{dx} \right)m(x)dx$$

$$= \frac{1}{2} \int_a^b \frac{d}{dx}\left(\frac{1}{S'(x)}\frac{df}{dx} \right)g(x)dx$$

$$= \frac{1}{2} \int_a^b \frac{1}{m(x)} \frac{d}{dx} \left(\frac{1}{S'(x)} \frac{df}{dx} \right) g(x) m(x) dx$$

$$= \int_a^b \mathscr{L} f(x) g(x) m(x) dx,$$

so this is indeed a *reversible* stationary distribution.

Example 3.29 (Which allele is the oldest?). Suppose that a population consists of two allelic types, a and A. We assume that one of the two alleles arose through mutation onto a background consisting entirely of the other type, since when there have been no further mutations. If the a-allele is currently at frequency p, what is the probability that it is the older allele?

The usual way to handle questions like this is to think of the model as one arising in the limit of very low mutation rates. If mutation rates are low then the process of allele frequencies consists of a sequence of excursions away from the boundaries. The a-allele is the oldest if as we trace backwards in time the allele frequency hits the boundary point 1 before it hits 0. Reversing with respect to the speed measure we see that this probability is the same as the probability that we hit 1 before 0 forwards in time. And (using Lemma 3.14) this in turn, as the mutation rates tend to zero, converges to the current frequency of the a-allele, that is p. For more details see, for example, Watterson (1977). $\square$

Chapter 4
More than Two Types

4.1 Multi-Dimensional Diffusion Models

So far we have considered only a very special case in which our population is classified into just two types. The frequencies are then characterised by a one-dimensional diffusion and one dimensional diffusions are, at least in principle, relatively straightforward to study. More generally, suppose that our population is classified into K different types. We're not going to develop the general theory of multidimensional diffusions, but let's see what happens in a special case.

Our starting point is a K-allele version of the Wright–Fisher model. The population configuration at any time can be described by a vector $(X_1, X_2, \ldots, X_K)$ where X_i is the number of genes of allelic type A_i and we assume that $X_1 + \cdots + X_K = N$. (Although only $K - 1$ components are necessary to specify the vector $(X_1, X_2, \ldots, X_K)$, it is sometimes convenient to retain all K.)

In the simplest case when all the alleles are selectively neutral and there is no mutation, we have

$$\mathbb{P}[Y_i \text{ genes of type } A_i \text{ at } t+1, i = 1, \ldots, K | X_j \text{ genes of type } A_j \text{ at } t,\ j = 1, \ldots, K]$$
$$= \frac{N!}{Y_1! Y_2! \cdots Y_K!} \psi_1^{Y_1} \psi_2^{Y_2} \cdots \psi_K^{Y_K}$$

where $\psi_i = \frac{X_i}{N}$ and $\sum_{i=1}^{K} Y_i = N$ (the probability is zero if this last condition is not satisfied).

We write $p_i(t) = X_i(t)/N$ and consider the increment $\delta p_i = p_i(t+1) - p_i(t)$. By 'pooling' all the alleles A_j for $j \neq i$ into a single class ('not A_i'), we recover the Wright–Fisher model for two alleles for which we already checked that

$$\mathbb{E}[\delta p_i] = 0, \quad \mathrm{var}(\delta p_i) = \frac{1}{N} p_i(1 - p_i) \quad \text{and} \quad \mathbb{E}[(\delta p_i)^k] = \mathcal{O}\left(\frac{1}{N^2}\right) \quad \forall k \geq 3.$$

To complete the picture we need the *covariances*, that is we must calculate, for $i \neq j$,

$$\mathbb{E}[\delta p_i \delta p_j] = \frac{1}{N^2} \mathbb{E}\left[(X_i(t+1) - X_i(t))(X_j(t+1) - X_j(t)) \big| \mathscr{F}_t \right]$$
$$= \frac{1}{N^2} \mathbb{E}\left[X_i(t+1) X_j(t+1) \big| \mathscr{F}_t \right] - p_i(t) p_j(t). \tag{4.1}$$

A. Etheridge, *Some Mathematical Models from Population Genetics*, Lecture Notes in Mathematics 2012, DOI 10.1007/978-3-642-16632-7_4,

Here $\mathscr{F}_t$ denotes all information about the process up until time t (more formally, it is the natural σ-field associated with $\{X_s\}_{s \le t}$). In the calculations below we write $\mathbb{E}_t$ for the corresponding conditional expectation. Now

$$\mathbb{E}_t[X_i(t+1)X_j(t+1)]$$
$$= \frac{1}{2}\left\{\mathbb{E}_t[(X_i(t+1)+X_j(t+1))^2] - \mathbb{E}_t[X_i(t+1)^2] - \mathbb{E}_t[X_j(t+1)^2]\right\}$$

and again 'pooling' genes of type A_i and A_j we reduce to a calculation for the binomial distribution, $Bin(N,p)$, for which we know

$$\mathbb{E}[X^2] = Np(1-p) + N^2 p^2.$$

This gives

$$\mathbb{E}_t[X_i(t+1)X_j(t+1)] = \frac{1}{2}\left\{N(p_i+p_j)(1-(p_i+p_j))+N^2(p_i+p_j)^2\right.$$
$$\left. -Np_i(1-p_i)-N^2 p_i^2 - Np_j(1-p_j)-N^2 p_j^2\right\}$$
$$= -Np_i p_j + N^2 p_i p_j$$

and substituting in (4.1) we obtain

$$\mathbb{E}[\delta p_i \delta p_j] = -\frac{1}{N}p_i p_j.$$

Now, just as in the two allele case, if we consider functions $f(p_1,\ldots,p_{K-1})$ (note that $p_K = 1 - \sum_{j=1}^{K-1} p_j$) and rescale time so that the inter-generation time is $1/N$ and let $N \to \infty$, then we can use Taylor's Theorem to identify the generator of the limiting model. This gives

$$\mathscr{L}f(p_1,\ldots,p_{K-1}) = \frac{1}{2}\sum_{i=1}^{K-1} p_i(1-p_i)\frac{\partial^2 f}{\partial p_i^2} - \sum_{1 \le i < j \le K-1} p_i p_j \frac{\partial^2 f}{\partial p_i \partial p_j}.$$

We can also add a mutation step. We did not do this explicitly in the two-allele Wright–Fisher model so let's be more explicit here. The idea is that for each offspring in each generation there is a small probability that it will not inherit the type of its parent, but rather it will mutate to another type. Suppose that with probability u_{ij} the offspring of a type A_i individual will (independently of one another) be type A_j. We will now have

$$\mathbb{E}[\delta p_i] = -\sum_{j \ne i} u_{ij} p_i + \sum_{j \ne i} u_{ji} p_j.$$

If we assume, as we did before, that mutation rates are very low (on the order of inverse population size), then writing $\beta_{ij} = Nu_{ij}$ we have

$$\mathbb{E}[\delta p_i] = \frac{1}{N}\left\{ -p_i \sum_j \beta_{ij} + \sum_j p_j \beta_{ji} \right\}.$$

The correction to $\mathbb{E}[(\delta p_i)^2]$ is of $\mathcal{O}(1/N^2)$. This gives the following lemma.

Lemma 4.1 (Multi-allele Wright–Fisher diffusion with mutation). *The generator of the K-allele Wright–Fisher diffusion with mutation is*

$$\frac{1}{2}\sum_{i=1}^{K-1} p_i(1-p_i)\frac{\partial^2 f}{\partial p_i^2} - \sum_{1\le i<j\le K-1} p_i p_j \frac{\partial^2 f}{\partial p_i \partial p_j} + \sum_{i=1}^{K-1}\left(-p_i \sum_j \beta_{ij} + \sum_j p_j \beta_{ji}\right)\frac{\partial f}{\partial p_i}.$$

If each $u_{ij} > 0$ for $i \ne j$ then the joint frequency of $A_1,\dots,A_{K-1}$ has a stationary distribution, but in general no closed form is known. It *is* known in the special case of symmetric parent-independent mutation.

Lemma 4.2. *Suppose that*

$$u_{ij} = \frac{u}{K-1}$$

so that the total mutation probability per gene per generation is u and it is equally likely to be a mutation to each of the other types. Then the corresponding K-allele Wright–Fisher diffusion with mutation has a stationary distribution with density

$$\psi(p_1,\dots,p_{K-1}) = \frac{\Gamma(K\varepsilon)}{(\Gamma(\varepsilon))^K}(p_1\cdots p_K)^{\varepsilon-1} \tag{4.2}$$

where $\varepsilon = \frac{2Nu}{K-1}$ and $p_K = 1 - p_1 - \dots - p_{K-1}$.

Proof. First note that in this case

$$-p_i \sum_j \beta_{ij} + \sum_j p_j \beta_{ji} = \frac{Nu}{K-1}(1-Kp_i).$$

Writing $\psi(p_1,\dots,p_{K-1})$ for the density of the stationary distribution and integrating by parts, exactly as we did to obtain (3.13) for the two-allele case, we find

$$\frac{1}{2}\sum_{i=1}^{K-1}\frac{\partial^2}{\partial p_i^2}(p_i(1-p_i)\psi(p_1,\dots p_{K-1})) - \sum_{1\le i<j\le K-1}\frac{\partial^2}{\partial p_i \partial p_j}(p_i p_j \psi(p_1,\dots,p_{K-1}))$$

$$-\sum_{i=1}^{K-1}\frac{\partial}{\partial p_i}\left(\frac{Nu}{K-1}(1-Kp_i)\psi(p_1,\dots,p_{K-1})\right)=0. \tag{4.3}$$

It is elementary (if tedious) to check that the expression in (4.2) solves this equation. $\square$

Notice that when $K = 2$, (4.3) becomes

$$0 = -\frac{1}{2}\frac{d}{dp}(\mu(1-2p)f(p)) + \frac{1}{2}\frac{d^2}{dp^2}(p(1-p)f(p))$$

where $\mu = 2Nu$ and (4.2) becomes

$$f(p) = \frac{\Gamma(2\mu)}{(\Gamma(\mu))^2}(p(1-p))^{\mu-1}$$

which is precisely the solution we found in (3.15), since in this notation $2v_1 = 2v_2 = \mu$.

In the 2-allele case we calculated the heterozygosity

$$H = \frac{4v_1 v_2}{(v_1 + v_2)(2(v_1 + v_2) + 1)}.$$

Substituting $v_1 = v_2 = Nu$ gives

$$H = \frac{2Nu}{4Nu + 1}.$$

Writing $\theta = 2Nu$ this gives

$$H = \frac{\theta}{2\theta + 1}.$$

Remark 4.3. This is the magic θ of Remark 2.18 but with 2 in place of 4 here because we have taken limits in a haploid population. To recover our previous θ we set $N = 2N_e$.

The expected *homozygosity*, F, which is the chance that a random sample of two genes is of the same allelic type is

$$F = 1 - H = \frac{\theta + 1}{2\theta + 1}.$$

For the K-allele model,

$$F = \sum_{i=1}^{K} \mathbb{E}[p_i^2] = K \int_0^1 p^2 p^{\varepsilon-1}(1-p)^{(K-1)\varepsilon-1}\frac{\Gamma(K\varepsilon)}{\Gamma(\varepsilon)\Gamma((K-1)\varepsilon)}dp$$

$$= K\frac{\Gamma((K-1)\varepsilon)\Gamma(\varepsilon+2)}{\Gamma(K\varepsilon+2)}\frac{\Gamma(K\varepsilon)}{\Gamma(\varepsilon)\Gamma((K-1)\varepsilon)}$$

$$= \frac{K\Gamma(\varepsilon+2)}{\Gamma(\varepsilon)}\frac{\Gamma(K\varepsilon)}{\Gamma(K\varepsilon+2)} = \frac{\varepsilon+1}{K\varepsilon+1}$$

and substituting $\varepsilon = \theta/(K-1)$ gives

$$F = \frac{\theta + K - 1}{K\theta + K - 1}. \tag{4.4}$$

Definition 4.4. The density (4.2) is called the *Dirichlet distribution*. It is usual to rearrange it and consider the sequence of allele frequencies in *decreasing* order:

$$p_{(1)} \geq p_{(2)} \geq \cdots \geq p_{(K)} \geq 0,$$

that is we look at the order statistics of $p_1, \ldots, p_K$. Their joint distribution is

$$f(p_{(1)}, \ldots, p_{(K)}) = \frac{K!\Gamma(K\varepsilon)}{\Gamma(\varepsilon)^K} \left(p_{(1)} \cdots p_{(K)}\right)^{\varepsilon - 1}.$$

Recall that the mutation model that led to this distribution was the symmetric parent-independent mutation model in which each individual mutates at the same rate to a type chosen uniformly from $\{1, \ldots, K\}$. If $K \to \infty$ this becomes the *infinitely many alleles model* (Definition 2.17). It is natural to ask whether we can find an analogue of the Dirichlet distribution for the stationary distribution of allele frequencies in the infinitely many alleles model. The answer, due to Kingman (1975, 1977), is yes.

4.2 The Poisson–Dirichlet and GEM Distributions

Kingman showed that, for every $j \geq 1$, the distribution of the first j order statistics of the Dirichlet distribution converges as $K \to \infty$ and called the corresponding limiting distribution the *Poisson–Dirichlet* distribution. In this section we shall try to understand why such a limit should exist. Direct manipulation of the Dirichlet distribution is difficult because of the linear dependence between the variables. However, it turns out that it can be represented in terms of *independent* Γ-random variables as follows.

Lemma 4.5. *Let* $Y_1, \ldots, Y_K$ *be independent positive random variables with probability density function*

$$g_\varepsilon(y) = \frac{y^{\varepsilon - 1} e^{-y}}{\Gamma(\varepsilon)}.$$

Then writing $Y = Y_1 + \cdots + Y_K$, *the vector* **p** *with components* $p_i = \frac{Y_i}{Y}$ *has the Dirichlet distribution and* Y *has a* Γ-*distribution with parameter* $K\varepsilon$. *Moreover,* **p** *is independent of* Y.

Proof. The proof of this claim is a simple change of variables,

$$(y_1, \ldots, y_K) \mapsto (p_1, \ldots, p_{K-1}, y).$$

In an obvious notation, $y_i = p_i y$ (and $p_K = 1 - \sum_{i=1}^{K-1} p_i$). Since the Y_i are independent,

$$f_{(p_1,\ldots,p_{K-1},Y)}(p_1,\ldots,p_{K-1},y) = f_{(Y_1,\ldots,Y_K)}(p_1 y,\ldots,p_K y) \left| \frac{\partial(y_1,\ldots,y_k)}{\partial(p_1,\ldots,p_{K-1},y)} \right|$$

$$= \prod_{i=1}^{K} \frac{(p_i y)^{\varepsilon-1} e^{-p_i y}}{\Gamma(\varepsilon)} \begin{vmatrix} y & & & p_1 \\ & y & & p_2 \\ & & \ddots & \\ & & & y \\ -y & -y & \cdots -y & p_K \end{vmatrix}$$

$$= \frac{1}{\Gamma(\varepsilon)^K} (p_1 \cdots p_K)^{\varepsilon-1} y^{K\varepsilon-K} e^{-y} \times y^{K-1} \sum_{i=1}^{K} p_i$$

$$= \frac{\Gamma(K\varepsilon)}{\Gamma(\varepsilon)^K} (p_1 \cdots p_K)^{\varepsilon-1} \frac{1}{\Gamma(K\varepsilon)} e^{-y} y^{K\varepsilon-1}$$

as required. $\square$

We now use this to find a representation of the Poisson–Dirichlet distribution (and in the process see why the name is natural). To do so we need a spatial analogue of the probability generating function of elementary probability.

Definition 4.6 (Probability generating functional). For a (possibly random) number of random points $\{Y_i\}_{i \in I}$ with each $Y_i \in (0, \infty)$ (say) we define the *probability generating functional* of $\{Y_i\}_{i \in I}$ by

$$G(\xi) = \mathbb{E}\left[\prod_{i \in I} \xi(Y_i) \right]$$

for any function $\xi : [0, \infty) \to \mathbb{R}$ for which the expectation exists.

If I is random, then we recover the probability generating function of I by choosing ξ to be constant.

Now choose the Y_i's to be independent Gamma random variables with parameter ε and consider the generating functional of $Y_1, \ldots, Y_K$. By independence,

$$G_K(\xi) = \left[\int_0^\infty \xi(u) \frac{u^{\varepsilon-1} e^{-u}}{\Gamma(\varepsilon)} du \right]^K .$$

Recall from (4.2) that $\varepsilon = 2Nu/(K-1)$ and $\theta = 2Nu$ so that $K\varepsilon \to \theta$ as $K \to \infty$. Now rewrite the term in square brackets using

$$\int_0^\infty \frac{u^{\varepsilon-1} e^{-u}}{\Gamma(\varepsilon)} du = 1 \quad \text{and} \quad \frac{\varepsilon}{\Gamma(\varepsilon+1)} = \frac{1}{\Gamma(\varepsilon)}$$

to obtain that

$$G_K(\varepsilon) = \left[1 - \varepsilon \int_0^\infty (1 - \xi(u)) \frac{u^{\varepsilon-1}}{\Gamma(\varepsilon+1)} e^{-u} du\right]^K$$

$$\to \exp\left(-\theta \int_0^\infty (1 - \xi(u)) u^{-1} e^{-u} du\right) \quad \text{as } K \to \infty.$$

The right hand side is the probability generating functional of a Poisson point process with intensity $\theta e^{-u}/u$, so in the limit as $K \to \infty$ the number of points in each interval $(a,b) \subseteq (0,\infty)$ is Poisson distributed with mean $\int_a^b (\theta e^{-u}/u) du$.

Now write $Y_{(1)} \geq Y_{(2)} \geq \cdots$ for the *ordered* points and $Y = Y_{(1)} + Y_{(2)} + \cdots$. Since $K\varepsilon \to \theta$ as $K \to \infty$, Y has a Gamma distribution with parameter θ.

Definition 4.7. The points $p_{(i)} = Y_{(i)}/Y$ have the *Poisson–Dirichlet* distribution.

The finite dimensional distributions of the $p_{(i)}$ are complicated, but those of the $Y_{(i)}$ are relatively straightforward. The density function of $Y_{(i)}$ is

$$\frac{\theta e^{-y}}{y} \frac{[\theta E_1(y)]^{i-1}}{(i-1)!} e^{-\theta E_1(y)}, \quad \text{for } y > 0,$$

where $E_1(y) = \int_y^\infty (e^{-u}/u) du$. Thus, for example, since $p_{(i)}$ is independent of Y (just as in Lemma 4.5)

$$\mathbb{E}[Y_{(i)}] = \mathbb{E}[p_{(i)}Y] = \mathbb{E}[p_{(i)}]\mathbb{E}[Y] = \theta \mathbb{E}[p_{(i)}]$$

gives

$$\mathbb{E}[p_{(i)}] = \frac{\theta^{i-1}}{(i-1)!} \int_0^\infty e^{-y}[E_1(y)]^{i-1} e^{-\theta E_1(y)} dy$$

which can be evaluated numerically.

In the Dirichlet distribution with K allelic types, the probability that there are alleles with frequencies in $(p_1, p_1 + dp_1), \ldots, (p_r, p_r + dp_r)$ is

$$\binom{K}{r} \frac{\Gamma(K\varepsilon)}{\Gamma(\varepsilon)^r \Gamma((K-r)\varepsilon)} (p_1 \cdots p_r)^{\varepsilon-1} \left(1 - \Sigma_1^r p_i\right)^{\varepsilon(K-r)-1} dp_1 \ldots dp_r$$

$$\to \theta^r (p_1 \cdots p_r)^{-1} \left(1 - \Sigma_1^r p_i\right)^{\theta-1} dp_1 \ldots dp_r \quad \text{as } K \to \infty. \tag{4.5}$$

In particular, taking $r = 1$, the probability that there is an allele with frequency in $(p, p+dp)$ under the limiting Poisson–Dirichlet distribution is $h(p)dp$ where

$$h(p) = \theta p^{-1}(1-p)^{\theta-1}.$$

Definition 4.8. The function $h(p) = \theta p^{-1}(1-p)^{\theta-1}$ is called the *frequency spectrum* of $\{p_{(i)}\}$.

The frequency spectrum allows us to calculate expressions of the form

$$\mathbb{E}\left[\sum_1^\infty f(p_{(i)})\right] = \int_0^1 f(p)h(p)dp$$

(provided this is finite). For example, taking $f(p_{(i)}) = p_{(i)}^2$ we calculate the *expected homozygosity*

$$F = \int_0^1 p^2 \theta p^{-1}(1-p)^{\theta-1}dp = \frac{1}{1+\theta}.$$

This is consistent with (4.4) as $K \to \infty$. Similarly, the expected number of alleles with frequencies in (a,b) is

$$\mathbb{E}\left[\sum_1^\infty 1_{(a,b)}(P_{(i)})\right] = \int_a^b \theta p^{-1}(1-p)^{\theta-1}dp,$$

and so on. This is the same θ cropping up again and again in our calculations.

The Poisson–Dirichlet distribution is not all that user-friendly, but remarkably a distribution obtained from it by 'size-biased' sampling is extremely elegant.

Example 4.9. Suppose that a gene is sampled at random from the population. What is the distribution of the frequency of alleles of the same type as the sampled individual?

The probability that the sampled allele has frequency in $[p, p+dp)$ is the probability that there *is* an allele with frequency in $[p, p+dp)$ *and* we choose it which is just

$$p \cdot \frac{\theta}{p}(1-p)^{\theta-1}dp = \theta(1-p)^{\theta-1}dp.$$

Now suppose that we remove all the individuals of this type and sample again from the remaining population. The chance that our new allele is at *relative* frequency r is just calculated by Bayes' rule. Let us write P_1 for the frequency of the first individual sampled and P_2 for the relative frequency of the second class. Then

$$\mathbb{P}[P_2 \in [r, r+dr)| P_1 \in [p, p+dp)]$$
$$= \frac{\mathbb{P}[\exists \text{ class with rel. freq} \in [r, r+dr) \text{ and sample from it and } P_1 \in [p, p+dp)]}{\mathbb{P}[P_1 \in [p, p+dp)]}$$
$$= \frac{\frac{\theta^2}{r(1-p)p}(1-p-r(1-p))^{\theta-1}pr\text{``}d(r(1-p))\text{''}dp}{\theta(1-p)^{\theta-1}dp} = \theta(1-r)^{\theta-1}dr,$$

where we have used (4.5) in the last line. In other words, P_2 has the same distribution as P_1. We can repeat this procedure and we find that the frequencies of the alleles picked in this way in our original population are

$$P_1, P_2(1 - P_1), P_3(1 - P_2)(1 - P_1), \ldots \qquad (4.6)$$

where the P_i are independent identically distributed random variables with density $\theta(1 - p)^{\theta - 1}, 0 < p < 1$.

Definition 4.10 (GEM distribution). The sequence of random variables in (4.6) are said to follow the *GEM distribution* after Griffiths, Engen and McCloskey (see Ewens (2004)).

All alleles in the population are eventually lost by the joint process of mutation and random drift and the probability that an allele at frequency p lives the longest of all current alleles is just p, so the GEM distribution can be thought of as allele frequencies when alleles are ordered according to their future persistence in the population. Reversibility arguments allow us to conclude that we have the same distribution if we order alleles by their age.

4.3 Ewens Sampling Formula

We now turn to one of the most famous results from mathematical population genetics. We continue to assume the infinitely many alleles model.

Definition 4.11 (Allele frequency spectrum). In a sample of size n, for $1 \le j \le n$ write $\alpha(j)$ for the number of alleles that occur exactly j times in our sample. The vector $(\alpha(1), \ldots, \alpha(n))$ is called the *allele frequency spectrum*.

This is illustrated in an example in Fig. 4.1.

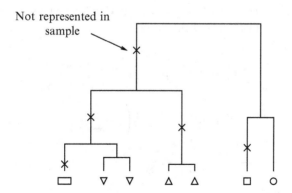

Fig. 4.1 The allele frequency spectrum. In this example, five mutations fall on the genealogical tree. Under the infinitely many alleles model, one of these mutations is not represented in the sample. The ancestral lineage of one individual has not experienced any mutations since the MRCA. In the notation of Definition 4.11, $\alpha(1) = 3$, $\alpha(2) = 2$, $\alpha(3) = \alpha(4) = \cdots = \alpha(7) = 0$ and so the allele frequency spectrum is $(3, 2, 0, 0, 0, 0, 0)$

Theorem 4.12 (Ewens sampling formula, Ewens 1972). *If the genealogy of our sample is determined by Kingman's coalescent, then under the infinitely many alleles model with mutation rate $\theta/2$ (in the coalescent timescale)*

$$\mathbb{P}\left[(\alpha(j))_{1\leq j\leq n} = (a_j)_{1\leq j\leq n}\right] = \frac{n!\theta^k}{a_1!\cdots a_n!1^{a_1}2^{a_2}\cdots n^{a_n}\theta_{(n)}}, \tag{4.7}$$

where $\sum_{j=1}^n a_j = k$ is the number of distinct alleles in the sample, $\sum_{j=1}^n ja_j = n$ and $\theta_{(n)} = \theta(\theta+1)\cdots(\theta+n-1)$.

Proof. There are many elegant derivations of this result (see, for example, Berestycki (2009) for one based on the 'Chinese Restaurant process'). Here we work directly with the Poisson–Dirichlet distribution.

First we show that if we take a sample of size n, then the probability that it falls into k distinct allelic types labelled $1, 2, \ldots, k$, say, with n_i individuals of type i for each i is

$$\frac{n!\theta^k}{n_1\cdots n_k\theta(\theta+1)\cdots(\theta+n-1)}. \tag{4.8}$$

To see this, recall that under the Poisson–Dirichlet distribution, the probability that there are points in $(p_1, p_1+dp_1), \ldots, (p_k, p_k+dp_k)$ is

$$\theta^k(p_1\cdots p_k)^{-1}\left(1 - \sum_1^k p_i\right)^{\theta-1} dp_1\ldots dp_k.$$

Given that such points exist, the probability that we see $n_1, \ldots, n_k$ copies of the corresponding alleles is the number of ways of assigning the n individuals in our sample to k classes of sizes $n_1, n_2, \ldots, n_k$ times the probability that the first n_1 are from class 1, the next n_2 from class 2 and so on. Combining these observations and integrating out over all choices of $p_1, \ldots, p_k$ gives

$$\frac{n!}{n_1!\cdots n_k!}\int_{\sum p_i \leq 1} p_1^{n_1}\cdots p_k^{n_k}\theta^k(p_1\cdots p_k)^{-1}\left(1-\sum_1^k p_i\right)^{\theta-1} dp_1\ldots dp_k$$

$$= \frac{n!}{n_1!\cdots n_k!}\theta^k\int p_1^{n_1-1}\cdots p_k^{n_k-1}\left(1-\sum_1^k p_i\right)^{\theta-1} dp_1\ldots dp_k$$

$$= \frac{n!}{n_1!\cdots n_k!}\theta^k\frac{\Gamma(n_1)\cdots\Gamma(n_k)\Gamma(\theta)}{\Gamma(n+\theta)}$$

$$= \frac{n!\theta^k}{n_1\cdots n_k}\frac{1}{\theta(\theta+1)\cdots(\theta+n-1)}$$

as required.

Now to obtain the corresponding probability for the allelic partition $(\alpha(j))_{1\leq j\leq n}$, we evaluate this for a vector $(n_1, \ldots, n_k)$ for which $n_i = j$ exactly a_j times and multiply by

$$\frac{1}{a_1!\cdots a_n!},$$

that is the number of ways of assigning the class sizes to k types divided by $k!$ (because we don't care about the labels attached to types), to obtain

$$\mathbb{P}[(\alpha(j))_{1\le j\le n} = (a_j)_{1\le j\le n}] = \frac{1}{a_1!\cdots a_n!}\frac{n!\theta^k}{n_1\cdots n_k}\frac{1}{\theta(\theta+1)\cdots(\theta+n-1)}$$

$$= \frac{n!\theta^k}{a_1!\cdots a_n!}\frac{1}{\prod_{j=1}^k j^{a_j}}\frac{1}{\theta_{(n)}}$$

where the last line follows because n_i is equal to j exactly a_j times.   □

Remark 4.13. This formula has some remarkable properties. Most importantly, note that if we condition on the number of distinct alleles in the sample being k, the distribution of $(\alpha(1),\ldots,\alpha(n))$ is *independent* of θ:

$$\mathbb{P}\left[(\alpha(j))_{1\le j\le n} = (a_j)_{1\le j\le n}\Big|\sum_{j=1}^n \alpha(j) = k\right] = \frac{\dfrac{n!\theta^k}{a_1!\cdots a_n!}\dfrac{1}{\prod_{j=1}^n j^{a_j}\,\theta_{(n)}}}{\displaystyle\sum_{(b_j):\sum b_j = k}\dfrac{n!\theta^k}{b_1!\cdots b_n!}\dfrac{1}{\prod_{j=1}^n j^{b_j}\,\theta_{(n)}}}$$

$$= \frac{1}{C_{n,j}}\frac{n!}{a_1!\cdots a_n!\prod j^{a_j}},$$

where the constant $C_{n,j}$ depends only on n and j. Thus it is possible to test the neutral theory – so the goodness of fit of the Kingman coalescent – without making *any* assumptions about N_e or θ.

This remark tells us that the number K of distinct allelic classes in our sample of size n is a sufficient statistic for θ, so how can we use K to estimate θ?

Lemma 4.14. *For a sample of size n, let K be the number of distinct alleles. Then*

$$\mathbb{E}[K] = 1 + \theta\sum_{j=2}^n \frac{1}{j+\theta-1}, \qquad \text{var}(K) = \theta\sum_{j=2}^n \frac{j-1}{(j+\theta-1)^2}.$$

In particular,

$$\mathbb{E}[K] \sim \theta\log n, \quad \text{var}(K) \sim \theta\log n \qquad \text{as } n\to\infty.$$

Remark 4.15. This suggests

$$\hat{\theta} = \frac{K}{\log n}$$

as an estimator for θ, but like Watterson's estimator (which was based on the infinitely many sites model), because $\mathbb{E}[K]$ grows only like $\log n$, convergence of $\hat{\theta}$ to θ is extremely slow.

Proof of Lemma 4.14. Consider the coalescent with mutation. As we trace backwards in time we think of ancestral lineages as being lost by mutation or by coalescence and K is then the number lost by mutation. Write

$$X_j = \begin{cases} 1 \text{ if the } (n-j+1)\text{th lineage to be lost is lost by mutation} \\ 0 \text{ otherwise.} \end{cases}$$

With this convention, X_j records whether the transition from j to $j-1$ ancestral lineages is by mutation $(X_j = 1)$ or coalescence $(X_j = 0)$. Then

$$\mathbb{P}[X_j = 1] = \frac{j\theta/2}{\binom{j}{2} + j\theta/2} = \frac{\theta}{j+\theta-1},$$

and so

$$\mathbb{E}[K] = \sum_{j=2}^{n} \mathbb{E}[X_j] = \sum_{j=2}^{n} \frac{\theta}{j+\theta-1}$$

and (since the X_j are independent)

$$\text{var}(K) = \sum_{j=2}^{n} \text{var}(X_i) = \sum_{j=2}^{n} \frac{\theta(j-1)}{(j+\theta-1)^2}.$$

$\square$

Chapter 5
Selection

5.1 Genetic Diversity

In Remark 2.18 we introduced the notion of *nucleotide diversity* – the proportion of nucleotides that differ between two randomly chosen sequences. Its expected value is $\theta = 4N_e\mu$ (for a diploid population) where μ is the mutation probability per base pair per individual per generation and N_e is the effective population size. The mutation rate can be estimated directly (or from the divergence between species with a known divergence time) and this gives an estimate of N_e (Barton et al. (2007), p.426). This approach yields $N_e \sim 10^6$ for *Drosophila melanogaster*, far lower than the actual (census) population size or indeed than the population size is likely to have been in the past. Moreover, although genetic variation is certainly higher in more abundant organisms, the relationship is rather weak. For example there's only about a factor of ten difference between *Drosophila melanogaster* and humans. Abundant species have much less genetic diversity than expected from the neutral theory, *something else is going on*.

Two explanations are generally put forward for why extremely abundant species like the bacteria *E. coli* or the fruitfly don't show correspondingly high levels of genetic diversity. One is *population bottlenecks* in which from time to time the population size is dramatically reduced (and then usually rather rapidly restored). Another is *natural selection*. Our aim in this chapter is to develop some of the mathematical tools necessary for studying these two effects. We primarily focus on selection.

5.2 Wright–Fisher Model with Selection

We are now going to consider populations in which individuals carrying different alleles have different fitnesses.

Definition 5.1. The *fitness of an individual* is the number of offspring that it leaves after one generation.

A. Etheridge, *Some Mathematical Models from Population Genetics*, Lecture Notes in Mathematics 2012, DOI 10.1007/978-3-642-16632-7_5,

The *fitness of a gene* is the number of copies it leaves after one generation.
The *fitness of an allele* is the average fitness of genes of that allelic type.

To model the evolution of the population forwards in time we extend the Wright–Fisher model to include selection.

Definition 5.2 (Wright–Fisher model with selection). In a panmictic, haploid population of constant size N, suppose that individuals are divided into two allelic types that we denote by a and A. If generation t consists of k individuals of type a and $N - k$ of type A then, according to the *Wright–Fisher model with selection*, the generation at time $t + 1$ is formed by sampling independently with replacement with

$$\mathbb{P}[a \text{ sampled}] = \frac{k(1+s)}{k(1+s) + N - k}, \quad \mathbb{P}[A \text{ sampled}] = \frac{N - k}{k(1+s) + N - k}. \quad (5.1)$$

The parameter s is called *selection coefficient*. We say that a, A have *relative fitness* $1 + s : 1$.

- If $s > 0$, a is said to be *beneficial*
- If $s < 0$, a is said to be *deleterious*

One way to think about the Wright–Fisher model is that each individual in generation t produces an effectively infinite pool of potential offspring, with proportions of different types dictated by (5.1), from which generation $(t + 1)$ is sampled.

We can also add mutation to this model. Suppose that during the reproductive step each type a individual from the pool mutates to A with probability μ_1 and each type A individual mutates to a with probability μ_2. Then the proportion of potential offspring which are type a after both selection *and* mutation is

$$\psi_k = \frac{k(1+s)(1-\mu_1)}{k(1+s) + N - k} + \frac{(N-k)\mu_2}{k(1+s) + N - k}. \quad (5.2)$$

Definition 5.3 (Wright–Fisher Model with selection and mutation). If there are k individuals of type a in generation t (and $N - k$ of type A), then under the Wright–Fisher model with selection and mutation, the number of type a individuals in generation $(t + 1)$ is $Bin(N, \psi_k)$ distributed where ψ_k is given by (5.2).

Evidently it is going to be extremely complicated to calculate anything explicitly for these models and so we pass to a diffusion approximation. As usual, time will be measured in units of N generations and we consider the *proportion* of type a individuals in the population. To obtain a non-trivial limit, we suppose that

$$\alpha = Ns, \quad \nu_1 = N\mu_1, \quad \nu_2 = N\mu_2. \quad (5.3)$$

Remark 5.4. It would be more usual to use the notation $\sigma = Ns$, but since σ was our notation for the variance of a one-dimensional diffusion we do not follow that convention.

Lemma 5.5. *As $N \to \infty$, the rescaled Wright–Fisher model converges to the one-dimensional diffusion with infinitesimal drift*

$$\mu(p) = \alpha p(1-p) - v_1 p + v_2(1-p),$$

and infinitesimal variance

$$\sigma^2(p) = p(1-p).$$

Proof. Let $\delta_t = \frac{1}{N}$ be the time between two generations (in rescaled time). As in the neutral case,

$$\mathbb{E}[(p_{1/N} - p)^k | p_0 = p] = \mathscr{O}\left(\frac{1}{N^2}\right) \qquad \text{for all } k \geq 3.$$

We must identify the infinitesimal mean and variance. If the current proportion of a alleles is p, the current *number* of type a individuals is $k \equiv Np$. We have

$$\mathbb{E}[(p_{1/N} - p) | p_0 = p] = \frac{1}{N}(N\psi_k - k)$$

and substituting

$$
\begin{aligned}
N\psi_k - k &= \frac{Nk(1+\frac{\alpha}{N})}{N+\frac{\alpha k}{N}}\left(1 - \frac{v_1}{N}\right) + \frac{N(N-k)}{N+\frac{\alpha k}{N}}\frac{v_2}{N} - k \\
&= \frac{1}{N+\frac{\alpha k}{N}}\left(Nk\left(1+\frac{\alpha}{N}\right)\left(1-\frac{v_1}{N}\right) + (N-k)v_2 - kN - \frac{\alpha k^2}{N}\right), \\
&= \frac{N}{N+\frac{\alpha k}{N}}\left(\frac{\alpha k}{N} - \frac{v_1 k}{N} + v_2 - \frac{v_2 k}{N} - \alpha\frac{k^2}{N^2} - \alpha\frac{v_1 k}{N^2}\right), \\
&= \alpha p - v_1 p + v_2 - v_2 p - \alpha p^2 + \mathscr{O}\left(\frac{1}{N}\right), \\
&= \alpha p(1-p) - v_1 p + v_2(1-p) + \mathscr{O}\left(\frac{1}{N}\right).
\end{aligned}
$$

The infinitesimal variance is easier. Since $\psi_k = k/N + \mathscr{O}(1/N)$,

$$
\begin{aligned}
\mathbb{E}\left[(p_{1/N} - p)^2 | p_0 = p\right] &= \frac{1}{N^2}N\psi_k(1-\psi_k) + \mathscr{O}\left(\frac{1}{N^2}\right), \\
&= \frac{1}{N}p(1-p) + \mathscr{O}\left(\frac{1}{N^2}\right).
\end{aligned}
$$

$\square$

Definition 5.6. We call the limiting diffusion in Lemma 5.5 the *weak selection limit* of the Wright–Fisher model with selection and mutation.

We can now use our knowledge of one-dimensional diffusions to explore this model. For example:

Lemma 5.7 (Fixation probabilities). *Suppose that there is no mutation ($v_1 = v_2 = 0$). If the initial proportion of a-alleles is p_0, the probability $p_{fix}(p_0)$ that the a-allele eventually fixes in the population (that is the diffusion is absorbed in $p \equiv 1$) is*

$$p_{fix}(p_0) = \begin{cases} \dfrac{1 - \exp(-2\alpha p_0)}{1 - \exp(-2\alpha)} & \text{if } \alpha \neq 0, \\ p_0 & \text{if } \alpha = 0. \end{cases} \tag{5.4}$$

Proof. Using Lemma 3.14, for $\alpha \neq 0$

$$p_{fix}(p_0) = \frac{S(p_0) - S(0)}{S(1) - S(0)}$$

$$= \frac{\displaystyle\int_0^{p_0} \exp\left(-\int_\eta^y \frac{2\mu(z)}{\sigma^2(z)} dz\right) dy}{\displaystyle\int_0^1 \exp\left(-\int_\eta^y \frac{2\mu(z)}{\sigma^2(z)} dz\right) dy}$$

$$= \frac{\displaystyle\int_0^{p_0} \exp(-2\alpha y) dy}{\displaystyle\int_0^1 \exp(-2\alpha y) dy}$$

$$= \frac{1 - \exp(-2\alpha p_0)}{1 - \exp(-2\alpha)}.$$

For $\alpha = 0$ the diffusion is already in natural scale and the result follows from Lemma 3.13. $\qquad\qquad\qquad\qquad\qquad\qquad\qquad\qquad\qquad\qquad\qquad\qquad\square$

If a selected allele arises through mutation in an otherwise neutral haploid population then its initial frequency is just $1/N$ and so, as a particular case of this lemma,

$$p_{fix}\left(\frac{1}{N}\right) = \frac{1 - \exp(-2\alpha/N)}{1 - \exp(-2\alpha)} = \frac{1 - \exp(-2s)}{1 - \exp(-2Ns)}. \tag{5.5}$$

Let's consider some special cases:

1. Deleterious alleles: $s < 0$. If $|s| \ll 1$ and $N|s| \gg 1$, $p_{fix}(1/N) \approx 2|s| \exp(-2N|s|)$. The fixation probability of a deleterious allele is exponentially small and it decreases with increasing population size.
2. Beneficial alleles: $s > 0$, $s \ll 1$, $Ns \gg 1$, then $p_{fix}(1/N) \approx 2s$, almost independent of population size.
3. Nearly neutral alleles: if $N|s| \ll 1$, then a is nearly neutral and $p_{fix}(1/N) \approx 1/N$.

In summary:

- Most alleles (beneficial or deleterious) are lost.
- Deleterious mutations are more likely to fix in small populations.
- Fitness differences that are too small to be measured in a laboratory ($|s| \ll 1$) can still play an important rôle in evolution (if $N|s| \gg 1$).

5.3 Selection in a Diploid Population

We have considered only so-called *genic selection*. There are many other more complex forms of selection. For example in diploid populations the fitness of an individual will typically depend on the combination of alleles that it carries at a particular locus. In this section we investigate how this affects the weak selection limit.

Suppose that an effectively infinite pool of gametes (cells containing one copy of each chromosome, see Sect. 5.6) fuse at random into diploid juveniles on which selection acts (for example selecting for different viabilities). From these juveniles we sample the adults that produce gametes for the next generation. As an example, suppose that we assign juvenile fitnesses as follows:

genotype	relative fitness
aa	$1 + s$
aA	$1 + hs$
AA	1

Definition 5.8. In this scheme, s is the selection coefficient of the *aa* homozygote and h is called the *degree of dominance* or the heterozygous effect.

For simplicity, we ignore mutation (it could be added just as before). If the proportion of a-alleles among the current population of gametes is p, then the juveniles have proportions

$$p_{aa} = p^2, \quad p_{aA} = 2p(1 - p), \quad p_{AA} = (1 - p)^2$$

and the new generation of adults (a finite number) will be sampled from a pool of juveniles with frequencies

$$p_{aa}(\text{adult}) = \frac{p^2(1 + s)}{\overline{w}}, \quad p_{aA}(\text{adult}) = \frac{2p(1 - p)(1 + hs)}{\overline{w}},$$

$$\text{and} \quad p_{AA}(\text{adult}) = \frac{(1 - p)^2}{\overline{w}},$$

where

$$\overline{w} = p^2(1+s) + 2p(1-p)(1+hs) + (1-p)^2 = 1 + sp^2 + 2hsp(1-p).$$

If the population size is finite, the resulting frequencies in the adult population are random. Denoting them $P_{aa}(\text{adult})$, $P_{aA}(\text{adult})$ and $P_{AA}(\text{adult})$, the new generation of gametes has frequency p' of a-alleles where

$$p' = P_{aa}(\text{adult}) + \frac{1}{2}P_{aA}(\text{adult}).$$

As before we suppose that the selection coefficient s is $\mathcal{O}(1/N)$ and then note that

$$
\begin{aligned}
\mathbb{E}[p' - p] &= \frac{p^2(1+s) + p(1-p)(1+hs)}{\overline{w}} - p \\
&= \frac{1}{1 + sp^2 + 2hsp(1-p)}\left\{p^2(1+s) + p(1-p)(1+hs)\right\} - p \\
&= \left(p + sp^2 + hsp(1-p)\right)\left(1 - sp^2 - 2hsp(1-p)\right) - p + \mathcal{O}(s^2) \\
&= sp^2 + hsp(1-p) - sp^3 - 2hsp^2(1-p) + \mathcal{O}(s^2) \\
&= sp(1-p)\left\{h + p(1-2h)\right\} + \mathcal{O}(s^2).
\end{aligned}
$$

We now write $\alpha = Ns$ and see that the drift in the diffusion that we obtain in the weak-selection limit will be

$$\mu(p) = \alpha p(1-p)\big(h + p(1-2h)\big).$$

If $h = 1/2$ then we obtain $\alpha p(1-p)/2$. (Note that this fits with the haploid case we considered before – with the parametrisation in our diploid model, the selective advantage of a is $s/2$, accounting for the additional factor of 2.) The variance turns out to be unchanged.

For this form of selection we see that the strength (and direction) of selection is frequency dependent. If the heterozygote is fitter than either homozygote, a situation known as *overdominance*, then selection works to maintain genetic variation. The classic example of this is sickle cell anaemia. When the heterozygote is less fit, we have *underdominance*. In the weak selection limit we can analyse the diffusion just as before.

5.4 The Ancestral Selection Graph

So far we have looked at the effect of selection in our forwards in time models. The next thing that we would like to understand is the effect of selection on genealogies. For the weak selection that we've looked at so far one can work with something

called the *ancestral selection graph* due to Neuhauser and Krone (1997), Krone and
Neuhauser (1997). To understand how it works, it is convenient to think about a
Moran model for the way that the population evolves forwards in time.

Definition 5.9 (Moran model with selection). In the Moran model for a haploid
population of size N, at rate $\binom{N}{2}$ a pair of individuals is selected at random, one dies,
the other reproduces. To incorporate selection, at an additional rate $\binom{N}{2}s$ a pair of
individuals is picked at random. Without loss of generality suppose that the a-allele
is beneficial (otherwise interchange the labels a and A). At these 'potential selective
events', if the chosen individuals are both of the same allelic type then nothing
happens; if one is a and the other is A, then the individual of type A dies and that of
type a reproduces (splits in two).

Remark 5.10. The parameter s plays the role of the selection coefficient that we had
before and will be $\mathscr{O}(1/N)$ in our weak selection limit. There are many other ways to
modify the neutral Moran model to incorporate selection. Here we have increased
the rate of reproduction events. One could equally assume that the reproduction
rate does not change, but if a pair consisting of one a and one A is chosen for
reproduction, then with probability $(1+s)/2$ it is the type a that reproduces and the
A that dies. The weak selection limit as $N \to \infty$ will be the same for both models.

Mutations are added as a Poisson process along the lineages, exactly as in the neutral
case (see Remark 2.29). Recall that we are already in the coalescent timescale and
so these will occur at the rescaled rates v_1 and v_2 of (5.3).

Lemma 5.11. *Suppose that* $\alpha = Ns$. *As* $N \to \infty$, *the Moran model with selection
and mutation converges to the same diffusion as the rescaled Wright–Fisher model
with selection and mutation.*

Proof. For a fixed N, the generator of the Moran model with selection is given by

$$
\mathscr{L}_N f(p) = \binom{N}{2} p(1-p) \left(f(p+\frac{1}{N}) - f(p) \right)
$$
$$
+ \binom{N}{2} p(1-p) \left(f(p-\frac{1}{N}) - f(p) \right)
$$
$$
+ N v_1 p \left(f(p-\frac{1}{N}) - f(p) \right) + N v_2 (1-p) \left(f(p+\frac{1}{N}) - f(p) \right)
$$
$$
+ 2s \binom{N}{2} p(1-p) \left(f(p+\frac{1}{N}) - f(p) \right). \tag{5.6}
$$

This is just the generator corresponding to the neutral Moran model with mutation plus an extra term corresponding to the selection. We take f to be three times

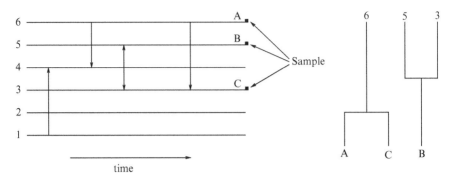

Fig. 5.1 Graphical representation of the Moran model with selection. The graphical representation mirrors that for the neutral model, but now additional 'double-headed' arrows, corresponding to 'potential selective events', are added to the picture. In order to decide the outcome of such an event, we need to know the types of the individuals at *both* ends of the arrow, immediately beforehand. On the right we see the branching and coalescing tree that tracks all 'potential ancestors' of individuals in our sample. We need to know the types of the individuals labelled 3 and 5 in order to determine which is the true ancestor of the individual labelled B

continuously differentiable. By Taylor's theorem, the last term on the right hand side of (5.6) is equal to $\alpha p(1-p)f'(p) + \mathcal{O}(1/N)$. Thus, as $N \longrightarrow \infty$, $\mathcal{L}_N f(p)$ converges to $\mathcal{L}f(p)$ where

$$\mathcal{L}f(p) = \frac{1}{2}p(1-p)f''(p) + (\nu_2 - (\nu_1 + \nu_2)p)f'(p) + \alpha p(1-p)f'(p). \quad (5.7)$$

$\square$

To understand the genealogy, we turn to the graphical representation of the Moran model, depicted in Fig. 5.1. This time we see *two* sorts of arrows. The 'neutral' arrows that we saw before are equally likely to point up or down. For such arrows (in the forwards in time Moran model) the individual at the 'tip' is replaced by a copy of the individual at the tail. But we also see *double headed* arrows corresponding to *potential selective events*. To resolve the outcome of a potential selective event corresponding to such an arrow we must know the types (before the event) of the individuals at *both* ends.

As we trace *backwards* in time, we see coalescence of ancestral lineages according to Kingman's coalescent exactly as before, but now we cannot resolve the double headed arrows. Instead, we trace *both* potential ancestors backwards in time. In this way we arrive at a *branching and coalescing* graph. It branches at potential selective events that hit a lineage already in the graph and it coalesces at the neutral events that hit two lineages in the graph. If eventually we get back to a single ancestral lineage, and if we can assign a type to that individual, then we can work our way back through the graph, deciding what actually took place at each potential selective event, and extract the coalescent tree that is the true genealogy of the sample. This will be best seen in an example, but first, for this procedure to work, we need to know that the graph will eventually collapse to a single lineage.

Now each individual is in $N-1$ pairs, each of which is hit by a potential selective event at rate $s = \alpha/N$. So as $N \to \infty$, each ancestral lineage will branch at rate α. On the other hand, each pair of lineages in the graph coalesces at rate 1. Writing L_t for the number of 'potential ancestral lineages', we see that $\{L_t\}_{t \geq 0}$ is a continuous time Markov chain with transition rates

$$k \mapsto k+1 \text{ at rate } \alpha k,$$
$$k \mapsto k-1 \text{ at rate } \binom{k}{2}.$$

Because the rate of decrease of the number of lineages is quadratic in k, whereas the rate of increase is only linear, in a finite (with probability one) random time the number of lineages in the graph will hit one for the first time. This single lineage corresponds to the *ultimate common ancestor*.

Definition 5.12. The graph traced out by the system of branching and coalescing lineages described above, stopped when the number of lineages hits one for the first time, is called the *ancestral selection graph*.

Note that the ultimate ancestor may have lived a very long time before the most recent common ancestor of our sample. To extract the true genealogy of the sample we must attach a type to the ultimate ancestor. If mutation rates between the two types are strictly positive (Krone and Neuhauser assume symmetric) then there is a stationary distribution for the Wright–Fisher diffusion that describes the frequency of types in the population. Indeed, in this particular case, the density m of the speed measure is given by

$$m(x) = Ce^{2\alpha x} x^{2v_2 - 1}(1-x)^{2v_1 - 1},$$

where C is a constant. In particular,

$$\int_0^1 m(x)dx < \infty.$$

Therefore

$$\psi(x)dx = \frac{m(x)}{\int_0^1 m(y)dy}dx$$

is a stationary measure for the diffusion. To decide the type of the ultimate ancestor, we sample from the stationary distribution and then work back through the tree to establish the genealogy of the sample. The promised example is in Fig. 5.2.

The problem with this approach is that unless α is very small, there is a huge proliferation of lineages and simulation of the ancestral selection graph becomes prohibitively computationally expensive. There are ways to improve this somewhat, but instead we consider a different approach. This approach will also allow us to move away from the assumption that our sample is random so that we will be able to describe the genealogy of a sample in which the *types* of individuals are known (see Etheridge and Griffiths (2009) for another approach, more akin to the ancestral selection graph, which also allows us to specify types in the sample). To understand the approach we return briefly to the basic Kingman coalescent.

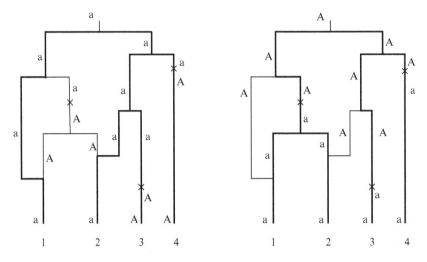

Fig. 5.2 Resolving the genealogy from the ancestral selection graph. For a particular realisation of the ancestral selection graph for a sample of size 4 we show the effect of assigning different types to the ultimate ancestor. The favoured allele is *a*. *Thick lines* represent the true genealogy. In this example the topology of the tree depends on the choice of type for the ultimate ancestor. Here the ultimate ancestor and the MRCA were the same (for both choices of type), but if we consider the subsample of $\{1,2\}$, then when the ultimate ancestor is type *a*, the MRCA is the same as the ultimate ancestor, but when the ultimate ancestor is type *A* the MRCA is more recent than the ultimate ancestor

5.5 Adding Structure to the Coalescent

We are going to proceed in two stages. One of the assumptions that we made in deriving Kingman's coalescent was that population size was constant. Our first task is to relax that assumption. Two key observations led us to the Kingman coalescent:

1. The probability that two individuals have a common parent is $1/N$ (and generations evolve independently);
2. for large N, the probability that three or more lineages merge in a single event or that two distinct pairs of individuals come together in a single generation is $\mathscr{O}(1/N^2)$.

Suppose that the population size is always large. We are going to measure it in units of size M (for example, M could be the current population size) and we shall also use M to rescale time. Write $N_M(t)$ for the population size t generations before the present. The chance that two lineages have *not* coalesced by time t is

$$\prod_{s=1}^{t}\left(1-\frac{1}{N_M(s)}\right)=\exp\left(\sum_{s=1}^{t}\log\left(1-\frac{1}{N_M(s)}\right)\right)$$

$$\approx \exp\left(-\sum_{s=1}^{t} \frac{1}{N_M(s)}\right), \quad \text{for } N_M(s) \text{ large.}$$

Now suppose that the rescaled and timechanged process $\frac{1}{M}N_M(Ms) \to \rho(s)$, as $M \to \infty$, for some 'nice' continuous function ρ. Then the probability that two lineages have not coalesced by time Mt is

$$\exp\left(-\sum_{s=1}^{Mt} \frac{1}{N(s)}\right) \approx \exp\left(-\sum_{s=1}^{Mt} \frac{1}{M\rho(s/M)}\right)$$

$$\approx \exp\left(-\int_0^t \frac{1}{\rho(s)}ds\right).$$

Exactly as in the derivation of Kingman's coalescent, at least provided $\rho(s) > 0$, we have that the probability of more than two lineages coalescing in a single generation or of coalescence of more than one pair of lineages in a single generation is $\mathcal{O}(1/M^2)$.

Lemma 5.13. *In the setting described above, as $M \to \infty$, if we measure time in units of size M, the number of lineages ancestral to a sample from the population alive at time t before the present converges to a pure death process with*

$$k \mapsto k-1 \text{ at instantaneous rate } \frac{1}{\rho(t)}\binom{k}{2}.$$

In other words, the genealogy of a sample is determined by a *timechange* of Kingman's coalescent. This works even when $\{\rho(s)\}_{s\geq 0}$ is stochastic, for example a diffusion (see Kaj and Krone (2003)). So we can relax our assumption of constant population size and still have a manageable coalescent model.

For our second extension we move away from the assumption of a *panmictic* population (in which every offspring is equally likely to choose any parent) and consider a *subdivided* population. Since we are interested in populations subject to selection, let us suppose that the population is subdivided into two allelic types labelled a and A subject to weak selection and mutation. Recall the Wright–Fisher model with selection and mutation of Definition 5.3: if there are currently k type a alleles, then the next generation is sampled from an infinite pool of potential offspring of which a proportion

$$\psi_k = \frac{k(1+s)(1-\mu_1)}{k(1+s)+N-k} + \frac{(N-k)\mu_2}{k(1+s)+N-k}$$

are of type a and the remainder are type A. Recall also, from the proof of Lemma 5.5, that

$$\psi_k = \frac{k}{N} + \mathcal{O}\left(\frac{1}{N}\right).$$

We now suppose that we sample n_1 individuals of type a and n_2 individuals of type A from among the offspring. We write $p = k/N$. Of the type a 'potential offspring' a proportion

$$\frac{\mu_2(1-p)}{p(1+s)(1-\mu_1)+\mu_2(1-p)} = \frac{v_2(1-p)}{Np} + \mathscr{O}\left(\frac{1}{N^2}\right)$$

arises (through mutation) from A parents. Similarly a proportion

$$\frac{v_1 p}{N(1-p)} + \mathscr{O}\left(\frac{1}{N^2}\right)$$

of the type A pool of potential offspring arises (through mutation) from type a parents. Since μ_1 and μ_2 are $\mathscr{O}(1/N)$, the probability that two ancestral lineages both arose through mutation (from either the same or different parents) in a single generation is $\mathscr{O}(1/N^2)$. The chance that two of the n_1 type a individuals have a common parent is

$$\binom{n_1}{2}\frac{1}{Np} + \mathscr{O}\left(\frac{1}{N^2}\right).$$

To see this, observe that in order to have a common parent, either neither arose through mutation or both did, and as noted above we can ignore the latter possibility and assume that they both arose from a type a parent. In the same way, the chance that two of the n_2 type A offspring share a common parent is

$$\binom{n_2}{2}\frac{1}{N(1-p)} + O\left(\frac{1}{N^2}\right).$$

Since a type a and a type A offspring can only have a common parent if one of them arose through mutation, we see that this has probability $\mathscr{O}(1/N^2)$.

Let us suppose that we knew the frequency $p(t)$ of type a parents in generation t before the present for each t and write $\rho(t) = p([Nt] + 1)$. Measuring time in units of N generations, up to an error of order $\mathscr{O}(1/N)$, a lineage ancestral to our sample that is currently of type a will jump to type A at instantaneous rate

$$\frac{v_2(1-\rho(t))}{\rho(t)}$$

(at rescaled time t). Similarly a single type A lineage will jump to type a at (approximately) instantaneous rate

$$\frac{v_1\rho(t)}{1-\rho(t)}.$$

Each pair of lineages that are currently type a will coalesce at instantaneous rate $1/\rho(t)$ (again up to an error of order $\mathscr{O}(1/N)$) and similarly two type A lineages coalesce at instantaneous rate $1/(1-\rho(t))$. Up to an error of order $\mathscr{O}(1/N)$, these are the only events that we see.

The difficulty now is to specify $p(t)$ and consequently $\rho(t)$. If the mutation rates μ_1 and μ_2 are strictly positive, then (since it is an irreducible aperiodic finite state space Markov chain) the Wright–Fisher model has a stationary distribution, but it is not a reversible stationary distribution and neither is it known explicitly, and so it is not possible to simply reverse the process with respect to the stationary measure. However, there is an explicit expression for the stationary distribution of the limiting Wright–Fisher diffusion and it is reasonable to guess that the reversed process converges to the time-reversal of the diffusion. A proof of joint convergence of the reversed Wright–Fisher model and the process describing the ancestral lineages of a sample from the population is going to be difficult, but if instead of the Wright–Fisher model one takes the Moran model with mutation and selection, then things are much simpler. The process of allele frequencies in the Moran model is just a birth and death process with a reversible stationary distribution and so convergence to the time-reversed diffusion is just a corollary of the convergence of the forwards in time model. Barton et al. (2004) show joint convergence of the time-reversed process of allele frequencies and the corresponding genealogies in a more general setting, essentially making rigorous the work of Darden et al. (1989). Consequently we have the following alternative to the ancestral selection graph.

Theorem 5.14 (Coalescent in a random background). *Measuring time in units of N generations, the distribution of the genealogy of a sample of n_1 type a individuals and n_2 type A individuals from a population undergoing weak genic selection and mutation can be obtained as follows. Let $\rho(t)$ be the time-reversal of the weak selection limit of Lemma 5.5 and write $n_1(t)$, $n_2(t)$ for the number of ancestral lineages of type a and of type A respectively at time t (that is $[Nt]$ generations) before the present. As $N \to \infty$, conditional on $\rho(t)$, the process of ancestral lineages evolves as follows:*

- $n_1 \mapsto n_1 - 1$ *at instantaneous rate* $\frac{1}{\rho(t)}\binom{n_1}{2}$
- $n_2 \mapsto n_2 - 1$ *at instantaneous rate* $\frac{1}{(1-\rho(t))}\binom{n_2}{2}$
- $\begin{cases} n_1 \mapsto n_1 - 1 \\ n_2 \mapsto n_2 + 1 \end{cases}$ *at instantaneous rate* $n_1 v_2 \frac{(1-\rho(t))}{\rho(t)}$
- $\begin{cases} n_1 \mapsto n_1 + 1 \\ n_2 \mapsto n_2 - 1 \end{cases}$ *at instantaneous rate* $n_2 v_1 \frac{\rho(t)}{(1-\rho(t))}$.

This approach has two advantages over the ancestral selection graph. First, we can specify the types of individuals in our sample. Second, the method is numerically practical even for strong selection. Changing the selection coefficient only changes the path $\rho(t)$.

5.6 Selective Sweeps

So far we have concentrated on the case where the rates of mutation between the two allelic types are both strictly positive. However, in applications, it is important to be able to remove this assumption. In particular, there is considerable interest in *selective sweeps*.

Definition 5.15. Suppose that an advantageous allele arising in an otherwise neutral population increases in frequency until the whole population carries it. Then it is said to have undergone a *selective sweep*.

We assume that there is no 'back-mutation' from the favoured to the less favoured type. There is no longer a reversible stationary distribution, but generally we are interested in the genealogy during the timecourse of the sweep and for that $\rho(t)$ can be specified by reversal with respect to the speed measure and conditioning on the backwards in time frequency of the selectively favoured a alleles hitting zero before one. This trick of reversing with respect to the speed measure goes back to Kimura. Mathematically it can be seen as reversing our previous model, with strictly positive mutation rates, and then taking a limit as mutation rates tend to zero (c.f. Example 3.29). This results in the equation

$$d\rho(t) = -\alpha\rho(t)(1 - \rho(t))\coth(\alpha(1 - \rho(t)))dt + \sqrt{\rho(t)(1 - \rho(t))}dW_t \quad (5.8)$$

where $\{W_t\}_{t \geq 0}$ is a standard Brownian motion. Suppose that we sample from the population at the time of completion of the sweep. Since the population is then all of type a, and there is no mutation, our ancestral lineages necessarily correspond exclusively to type a individuals and the genealogy is as in Lemma 5.13 with $\rho(t)$ determined by (5.8).

The difficulty is that typically we don't know which loci on the genome are undergoing selection and indeed that is what we'd like to find out. Moreover, if an allele at a particular locus is fixed, there is no variability in a sample from that locus from which to infer genealogical relationships. To identify loci that have been subject to selection, one looks at sites that are known (or at least believed) to be neutral. If the pattern of variation at such a site is not what we'd expect under the neutral model then we suspect that something else is happening at another locus on the same chromosome. In diploid populations we can even say that it has to be at a 'nearby' locus. To see why we need a little more biology.

In a diploid population, such as our own, in which chromosomes are carried in pairs (leaving aside the X and Y chromosomes) it is not the case that chromosomes are passed down as indivisible blocks. Although we inherit one of each pair from our mother and one from our father, the chromosome that I passed down to my daughter is *not* an exact copy of one of my chromosomes. Instead it is a mosaïc of my two chromosomes. This is due to a process called *recombination*. The simplest sort of recombination is a so-called *crossover event*. The cartoon in Fig. 5.3 gives an idea of what is going on. During reproduction each parent produces a large number of *germ cells*, each carrying a copy of both the chromosomes in the pair. During *meiosis* the genome of a diploid germ cell undergoes DNA replication followed by two rounds of division which results in four haploid cells or *gametes*. Each of these contains just one complete set of chromosomes. During this process the pairs of chromosomes can randomly exchange segments of genetic information as in our cartoon. Thus all four combinations, AC, AD, BC and BD can be represented among the gametes produced by this parent. Two gametes, one chosen at random from each

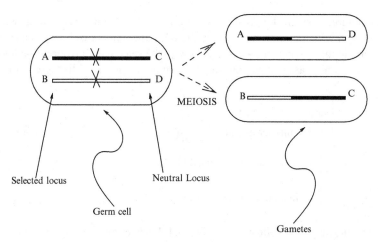

Fig. 5.3 Recombination. This cartoon represents a 'crossover event'. A neutral allele that was associated with an A at the selected locus in the parental chromosome can be passed on to the offspring in association with B

parent, fuse to form an offspring. In particular, as a result of such crossover events, a neutral allele that was associated with an A at the selected locus in the parental chromosome can be passed on to the offspring in association with B.

Now think of the population at the neutral locus as being subdivided into backgrounds a and A. As we trace backwards in time we see coalescence within each background exactly as before, but now a lineage can move between backgrounds not due to mutation, but due to recombination. We suppose that the probability that a gamete resulted from a crossover event is R. We can proceed by analogy with the arguments that lead to Theorem 5.14. Whereas a mutation of a type A to a type a individual occurred with probability μ_2 per generation, a crossover event affects a type A individual with probability R. However, this crossover will only result in a change of type at the selected locus if it happens in association with a chromosome carrying type a at the selected locus and, since we formed our diploid through a random fusion of gametes, this happens with probability p. Thus we must take Rp in place of μ_2. Similarly, in place of μ_1 we take $R(1-p)$. Thus, tracing backwards in time, the probability that a lineage that is currently associated with a type a allele at the selected locus arose through recombination with an individual in background A is

$$\frac{Rp(1-p)}{p} = R(1-p).$$

Similarly, the probability that a lineage currently in background A arose through recombination with an individual in background a is

$$\frac{Rp(1-p)}{1-p} = Rp.$$

Recombination rates are low and so writing $r = NR$ and mimicking what we did for the model with mutation we see that, denoting the number of lineages ancestral to the sample that are currently associated with the selectively favoured a-allele by n_1, and the number associated with the A-allele by n_2, ancestral lineages evolve as follows:

- $n_1 \mapsto n_1 - 1$ at instantaneous rate $\frac{1}{\rho(t)} \binom{n_1}{2}$
- $n_2 \mapsto n_2 - 1$ at instantaneous rate $\frac{1}{(1-\rho(t))} \binom{n_2}{2}$
- $\begin{cases} n_1 \mapsto n_1 - 1 \\ n_2 \mapsto n_2 + 1 \end{cases}$ at instantaneous rate $n_1 r (1 - \rho(t))$
- $\begin{cases} n_1 \mapsto n_1 + 1 \\ n_2 \mapsto n_2 - 1 \end{cases}$ at instantaneous rate $n_2 r \rho(t)$.

A typical genealogy is illustrated in Fig. 5.4. If selection is strong, then the time-course of the sweep is of order $\mathcal{O}(\log \alpha / \alpha) = \mathcal{O}(\log(Ns)/(Ns))$ units of time (in the coalescent scaling). If we are to see a non-trivial pattern of variation at the neutral locus, then the scaled recombination rate, r, between the selected and the neutral locus must be $\mathcal{O}(\alpha / \log \alpha)$. Here, by non-trivial we mean that the presence of the sweep affects the pattern of variation, but is not felt so strongly that there is no variation at all at the neutral locus. With a rate of this order, a given ancestral lineage has strictly

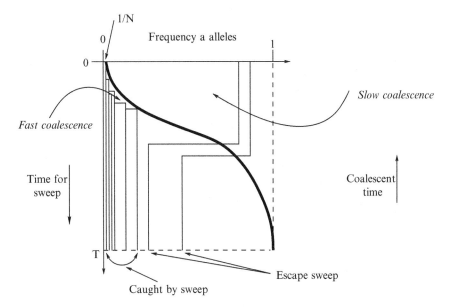

Fig. 5.4 The effect of a selective sweep on the genealogy at a linked neutral locus. The bold curve depicts the frequency of the favoured allele as it sweeps to fixation. Lineages in the selected background coalesce quickly at times close to the beginning of the sweep, leading to a characteristic 'star shaped' genealogy. Lineages that recombine into the unfavoured background are unlikely to coalesce, favouring singletons (neutral alleles that are represented only once) in the sample

positive probability of recombining during the course of the sweep. On the other hand, during most of the sweep the frequency of *a* alleles is either very close to 0 or very close to 1. As a result it is very rare to see a lineage move out of background *a* and later move *back* into *a* since, by the time a lineage has recombined into the *A* background, the *a*-alleles are scarce and so even if the lineage experiences a recombination event it will not be in association with a type *a* individual. Moreover, it is rare to see coalescence of two lineages in background *A* since, by the time two lineages have moved to background *A*, the frequency $1 - p(t)$ of *A*-individuals is $\mathscr{O}(1)$ and so the rate of coalescence within the *A* background is slow ($\mathscr{O}(1)$) compared to the duration of the sweep.

Let us suppose that every individual in the population at the time of the origin of the sweep has a different type at the neutral locus. Then the above heuristics tell us that a typical pattern of types in our sample will be one large family, derived from the chromosome on which the favoured mutation originally arose, and a number of singletons (corresponding to lineages which recombine out of background *a* during the sweep). In Fig. 5.4, the individuals that are caught by the sweep form a single family while those that escape the sweep are singletons. This pattern is regarded as a 'signature' of a selective sweep. One can check that the probability of seeing a pattern different from this is $\mathscr{O}(1/\log(Ns))$. More refined results can be found in Durrett and Schweinsberg (2004), Schweinsberg and Durrett (2005) and Etheridge et al. (2006). See also Barton (1998).

We have assumed that we are sampling from the neutral locus at the time of fixation of the favoured allele. In practice this is never the case. The signature of selection at the neutral locus will be masked by a period of Kingman coalescence of lineages as we trace back from the present to the time of completion of the sweep.

We have established the shape of the genealogical trees relating a sample from the neutral locus, but what's happening to allele frequencies of the neutral locus forwards in time? Figure 5.5 provides a cartoon. The allele at the neutral locus on the chromosome on which the favourable allele originally arose will receive a 'boost' in frequency which we have denoted by *u*.

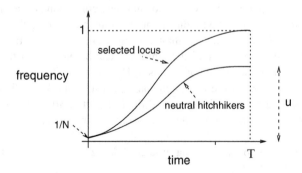

Fig. 5.5 Genetic hitchhiking. During a selective sweep, allele frequencies can become distorted at linked neutral loci. In particular, the neutral allele carried by the chromosome on which the favoured mutation originally arose receives a boost in frequency, denoted here by *u*

Definition 5.16. We call this boost in frequency at the neutral locus *genetic hitchhiking.*

Given the random value u, according to the approximation above, if there are n ancestral lineages at time T (the time of fixation of the favourable allele), then as we trace back through the sweep a $Bin(n,u)$-distributed set of them will coalesce to trace back to the founder of the sweep and the remainder will be unaffected during the sweep.

Because we are considering very rapid sweeps (corresponding to $Ns \gg 1$), there will rarely be sufficient mutations to resolve the details of the genealogy during the sweep and so it is equivalent to think of the sweep as leading to an instantaneous coalescence of a binomial number of lineages. If we think of a neutral site as being affected by a series of such selective sweeps, then the genealogy there will be represented as a coalescent with *multiple* mergers, see Remark 5.25. (In this context, 'multiple' means more than two.) Such coalescents are our next topic.

5.7 Coalescents with Multiple Mergers

The processes now known as Λ-coalescents were introduced independently by Donnelly and Kurtz (1999), Pitman (1999) and Sagitov (1999). Like Kingman's coalescent they take their values among partitions of $\mathbb{N}$ and their laws can be prescribed by specifying the restriction to partitions of $[k] = \{1, 2, \ldots, k\}$ for each $k \in \mathbb{N}$.

Definition 5.17 (Λ-**coalescent**). A Λ-coalescent, $\{\pi(t)\}_{t \geq 0}$, is a Markov process taking its values among partitions of $\mathbb{N}$ with the property that for each k, the restriction to $[k]$, $\{\pi_k(t)\}_{t \geq 0}$, is also a Markov process and if there are currently n blocks in $\pi_k(t)$ then each transition involving j of the blocks merging into one happens at rate $\beta_{n,j}$ (which is *independent* of k) and these are the only possible transitions.

Generally $\pi(0)$ is taken to be the partition all of whose blocks are singletons. For our purposes the Λ-coalescent describes the ancestry of a population whose individuals are labelled by $\mathbb{N}$. Each block in the partition at time t corresponds to a single ancestor at time t before the present with the elements of the block being the descendants of that ancestor. Whereas for the Kingman coalescent the only transitions are mergers of pairs of blocks, for the Λ-coalescent there can be mergers of three or more blocks. The key point is that the conditions in Definition 5.17 ensure the sampling consistency that we saw for the Kingman coalescent in Remark 2.4: the coalescent obtained by taking a sample of size k_1 and then restricting to a subsample of size $k_2 < k_1$ has the same distribution as the coalescent obtained by simply starting from a sample of size k_2.

If such a process is to exist, the parameters $\{\beta_{n,j}, 2 \leq j \leq n\}$ cannot be chosen arbitrarily.

Theorem 5.18 (Pitman 1999). *The Λ-coalescent of Definition 5.17 exists if and only if there is a finite measure Λ on $[0,1]$ for which*

$$\beta_{n,j} = \int_{[0,1]} u^{j-2}(1-u)^{n-j}\Lambda(du). \tag{5.9}$$

Remark 5.19 (Recovering Kingman's coalescent). Notice that if $\Lambda = \delta_0$, the point mass on zero, then $\beta_{n,j}$ vanishes unless $j = 2$ (corresponding to two blocks, or ancestral lineages, merging), in which case it is one, so we recover Kingman's coalescent.

More generally one can consider coalescents with *simultaneous* multiple collisions. Such coalescents were obtained as the limit of the ancestral processes of properly scaled population models (in much the same way as the Kingman coalescent arises as the rescaled ancestral process corresponding to the Wright–Fisher model) by Möhle and Sagitov (2001). They were able to identify all coalescents arising as limits of genealogies of populations with exchangeable reproduction mechanisms. Schweinsberg (2000) independently obtained the same class of coalescents (without a passage to the limit) and characterised the possible rates of mergers of ancestral lineages in terms of a single measure Ξ on the infinite simplex

$$\Delta = \left\{ (x_1, x_2, \ldots) : x_1 \geq x_2 \geq \ldots \geq 0, \sum_{i=1}^{\infty} x_i \leq 1 \right\}.$$

The resulting coalescents are now known as Ξ-coalescents.

Definition 5.20 (Ξ-coalescent). Let Ξ be a finite measure on Δ and write $\Xi = \Xi_0 + a\delta_0$ where Ξ_0 has no atom at zero. The Ξ-coalescent (with measure Ξ) is a process $\{\pi(t)\}_{t \geq 0}$ taking values among partitions on $\mathbb{N}$ with the property that for each k, the restriction to partitions of $[k]$ is also a Markov process. Let ζ be a partition with n blocks and let η be a partition obtained by merging disjoint groups of blocks of ζ. If there are $k_1, \ldots, k_r$ blocks in each group, where $k_i \geq 2$ for $i = 1, \ldots, r$, so that η has $n - \sum_{i=1}^{r} k_i + r$ blocks, then writing $s = n - \sum_{i=1}^{r} k_i$, the rate of transition from ζ to η is

$$\int_{\Delta} \left(\sum_{l=0}^{s} \sum_{i_1 \neq i_2 \neq \ldots \neq i_{r+l}} \binom{s}{l} x_{i_1}^{k_1} \cdots x_{i_r}^{k_r} x_{i_{r+1}} \cdots x_{i_{r+l}} \left(1 - \sum_{j=1}^{\infty} x_j \right)^{s-l} \right) \frac{\Xi_0(dx)}{\sum_{j=1}^{\infty} x_j^2}$$
$$+ a\mathbf{1}_{\{r=1, k_1=2\}}, \tag{5.10}$$

and these are the only possible transitions.

The rates (5.10) will be explained in Remark 5.22. For most of what follows we shall restrict our attention to Λ-coalescents, not least because this greatly simplifies notation, but it should be clear that many ideas carry over to this more general setting.

The Λ-coalescent specifies the genealogy of a sample from a forwards in time population model introduced by Bertoin and Le Gall (2003) that we shall call the Λ-Fleming-Viot process. This model is implicit in Donnelly and Kurtz (1999). See also Birkner et al. (2005) and Berestycki et al. (2007) for explicit simultaneous constructions of the Λ-Fleming-Viot process and its genealogical trees for certain classes of measure Λ.

The Λ-Fleming-Viot process takes its values among probability measures on $[0,1]$. We will describe it in terms of its generator, $\mathscr{R}$, acting on functions of the form

$$G(\rho) = \int f(x_1,\ldots,x_n)\rho(dx_n)\ldots\rho(dx_1), \qquad (5.11)$$

where $n \in \mathbb{N}$ and $f : [0,1]^n \to \mathbb{R}$ is measurable and bounded. First we need some notation. If $x = (x_1,\ldots,x_n) \in [0,1]^n$ and $J \subseteq \{1,\ldots,n\}$ we write

$$x_i^J = x_{\min J} \text{ if } i \in J, \text{ and } x_i^J = x_i \text{ if } i \notin J, i = 1,\ldots,n. \qquad (5.12)$$

In words, for each coordinate whose index is in J, we substitute the value of the coordinate with the lowest index from J.

Definition 5.21 (Λ-Fleming-Viot process). Let Λ be a finite measure on $[0,1]$. The Λ-Fleming-Viot process has generator

$$\mathscr{R}G(\rho) = \sum_{J\subseteq\{1,\ldots,n\},|J|\geq 2} \beta_{n,|J|} \int \left(f(x_1^J,\ldots,x_n^J) - f(x_1,\ldots,x_n)\right)\rho(dx_n)\ldots\rho(dx_1), \qquad (5.13)$$

where $\beta_{n,j}$ is defined in (5.9). When $\Lambda(\{0\}) = 0$ (so there is no Kingman component), this can also be written

$$\mathscr{R}G(\rho) = \int_{(0,1]}\int_{[0,1]} \left(G((1-u)\rho + u\delta_k) - G(\rho)\right)\rho(dk)u^{-2}\Lambda(du). \qquad (5.14)$$

The Λ-Fleming-Viot process is most easily understood in the case $\Lambda(\{0\}) = 0$. In that case one can think of it as follows. The population at time t is described by a probability measure $\rho(t)$ on the type space $[0,1]$. Take a Poisson point process on $\mathbb{R}_+ \times (0,1]$ with intensity measure $dt \otimes u^{-2}\Lambda(du)$ which picks times and sizes of jumps for our population process. At a jump time t with corresponding jump size u, a proportion u of the population is killed and replaced by offspring of an individual chosen at random from $\rho(t-)$. Thus

$$\rho(t) = (1-u)\rho(t-) + u\delta_x$$

where $x \in [0,1]$ is chosen according to $\rho(t-)$. The duality with the Λ-coalescent is evident. To construct the genealogy of a sample from such a population, suppose that there are currently k ancestral lineages. We trace backwards in time until we first encounter a point of our Poisson process. Suppose that the jump size that this specifies is u. Then for each lineage, independently, we flip a coin which shows heads with probability u. All lineages with a head merge into a common ancestor and the process continues.

Remark 5.22 (Poissonian construction of Ξ-coalescents.). A Poisson construction also allows us to understand the rates (5.10) in the Ξ-coalescent. Suppose that $a = 0$ and take a Poisson point process, Π, on $[0,\infty) \times \Delta$ with intensity

$$dt \otimes \frac{\Xi_0(dx)}{\sum_{i=1}^{\infty} x_i^2}.$$

If $(t,\mathbf{x}) \in \Pi$, then tracing backwards in time, each block of the partition (equivalently each ancestral lineage) alive immediately before the event, independently of the others, chooses colour i with probability x_i and stays the same colour with probability $1 - \sum_{i=1}^{\infty} x_i$. During the coalescence event, blocks of the same colour merge.

To understand the rate in (5.10), notice that a block is not changed during the event if either it didn't change colour, or if it was the only block to choose a particular colour. The 'l' in the formula corresponds to the blocks that are not involved in the merger, but were nonetheless coloured, and the coefficient $\binom{s}{l}$ gives the number of ways of choosing those blocks. If there are l such blocks then $r + l$ different colours were chosen by at least one block (leading to the sum over $i_1 \neq \ldots \neq i_{r+l}$). For each choice of colours and for each $m = 1, \ldots, r$, the term $x_{i_m}^{k_m}$ is the probability that k_m blocks are coloured i_m and merge. Finally, $(1 - \sum_{i=1}^{\infty} x_i)^{s-l}$ is the probability that none of the remaining $s - l$ blocks is coloured.

Perhaps the most extensively studied class of Λ-coalescents arise when the measure Λ is a Beta-measure.

Definition 5.23 (Beta-coalescents). The special class of Λ-coalescents for which

$$\Lambda(dx) = \frac{1}{\Gamma(2-\alpha)\Gamma(\alpha)} x^{1-\alpha}(1-x)^{\alpha-1} dx, \qquad \alpha \in (0,2),$$

are called the *Beta-coalescents*.

For each fixed $\alpha \in (0,2)$, the corresponding Λ-Fleming-Viot process arises as a timechange of the process which records proportions of different types in a population evolving according to an α-stable branching process in which individuals inherit the type of their parent (see Birkner et al. (2005)). This renders them particularly amenable to analytic study. For $1 \leq \alpha < 2$, they were obtained by Schweinsberg (2003) from populations in which individual offspring distributions have heavy tails. To see how this works, suppose that in a population of size N each individual (independently) produces a random number of juvenile offspring according to a distribution ψ. Density-dependent population regulation operates so that exactly N juveniles, sampled at random, survive to maturity. We assume that the mean number m of juveniles produced by each individual is greater than one so that as $N \to \infty$ the probability that there are greater than N juveniles from which to select the next generation tends to one. We are really looking at a Cannings model. Writing X for the number of juvenile offspring of a given parent, suppose that

$$\mathbb{P}[X > k] \sim \frac{C}{k^\alpha} \quad \text{as } k \to \infty$$

for a suitable constant C and for $\alpha > 0$. Let us write v_1 for the number of these juveniles that survive to maturity and

$$c_N = \frac{\mathbb{E}[v_1(v_1 - 1)]}{N - 1}.$$

This is precisely the quantity c_N of Sect. 2.2. Since v_1 is certainly no more than $\min(X, N)$, $\mathbb{E}[v_1^2] \leq \mathbb{E}[\min(X, N)^2]$ from which c_N is at most $\mathscr{O}(N^{1-\alpha})$. In particular, for $\alpha > 1$, $c_N \to 0$ as $N \to \infty$. For $\alpha \geq 2$ the conditions of Lemma 2.13 are satisfied and we recover a Kingman coalescent. For $1 < \alpha < 2$, the condition (2.4) fails. However, writing v_2 for the number of mature offspring of a second individual, it is easy to check that

$$\frac{\mathbb{E}[v_1(v_1 - 1)v_2(v_2 - 1)]}{N^2 c_N} \to 0 \quad \text{as } N \to \infty. \tag{5.15}$$

This guarantees that in times of $\mathscr{O}(1/c_N)$ generations, we do not see *simultaneous* mergers of two different pairs of ancestral lineages, so that we can expect a Λ-coalescent in the limit as $N \to \infty$. To a good approximation, for this range of α, we can ignore the possibility of seeing more than one very large family (of size $\mathscr{O}(N)$) in a single generation. In the generations when we *do* see a large family, we can expect the remaining $N - 1$ parents to produce a total of about $m(N - 1)$ juveniles. Conditional on a large family, whose size we denote by X, in order to see a jump of size u in the types of juveniles, we need

$$\frac{X}{X + (N - 1)m} \geq u.$$

Now conditioning on there being a large family we find

$$\mathbb{P}\left[\frac{X}{X + (N-1)m} \geq u \,\middle|\, X \geq \varepsilon N\right]$$
$$= \mathbb{P}\left[X \geq (N-1)m\frac{u}{1-u} \,\middle|\, X \geq \varepsilon N\right] \sim C'\frac{(1-u)^\alpha}{u^\alpha},$$

(for a suitable constant C') and the right hand side is (up to a constant) the Beta-measure of $[u, 1]$.

The case $\alpha = 1$ requires a different argument, for which we refer to Schweinsberg (2003). When $0 < \alpha < 1$ we no longer have that $c_n \to 0$. Moreover, condition (5.15) fails and in the limit we obtain a coalescent with *simultaneous* multiple mergers, that is, a Ξ-coalescent.

Remark 5.24. Notice that the existence of the Λ-Fleming-Viot process corresponding to a Beta-measure as above with $1 < \alpha < 2$ is slightly delicate. It depends upon cancellation of small 'upwards' and 'downwards' jumps. To see this, suppose that the population is divided into a and A alleles and that the current frequency of a-alleles is p. Then the generator acting on a function $g(p)$ takes the form

$$
\mathscr{R}g(p) = \int_0^1 \left(p\{g(p+u(1-p))-g(p)\} \right.
$$
$$
\left. +(1-p)\{g(p(1-u))-g(p)\}\right)\frac{1}{u^2}u^{1-\alpha}(1-u)^{\alpha-1}du
$$

with the first term in the integrand corresponding to 'upwards' jumps and the second to 'downwards' jumps. Expanding g in a Taylor series, we see that the terms of order 1 and u cancel, leaving an integrand of order u^2 to counteract the $1/u^2$ in the Λ-measure. We return to this point in Remark 6.19.

Remark 5.25 (Gillespie's pseudohitchhiking model). A basic observation, which we already made at the beginning of Chap. 5 and to which we shall return in Sect. 6.5, is that genetic diversity is orders of magnitude lower than expected from census numbers and genetic drift alone. One factor that may contribute to this is repeated selective sweeps. Gillespie proposed a model in which advantageous mutations arise in a population according to a Poisson arrival process. He supposes that selection is strong so that for those favourable mutations that become established, not only is fixation essentially certain, but also the corresponding selective sweeps are very rapid and can be viewed as (almost) instantaneous. Moreover (because such strongly selected mutations should be rare) sweeps corresponding to different mutations are assumed not to overlap. He examines the effect on a linked neutral locus of such repeated substitutions. In this *pseudohitchhiking model*, the genealogy of a sample from the neutral locus can be approximated by a Λ-coalescent in which $dt \otimes u^{-2}\Lambda(du)$ is determined by the rate of substitutions (that is selective sweeps) and the distribution of the random variable u, introduced above Definition 5.16, that quantifies the 'hitchhiking effect' for each sweep. A more accurate approximation for large sample sizes is provided by a Ξ-coalescent (see Durrett and Schweinsberg (2005)).

The mathematical theory of Λ-coalescents is developing extremely quickly. Here we have barely scratched the surface. For a nice survey we refer to Berestycki (2009). In biology, both Λ and Ξ-coalescents have attracted interest as models for the genealogies of samples from certain marine organisms, in which the distribution of offspring number among individuals is highly skewed, see for example Eldon and Wakeley (2006), Sargsyan and Wakeley (2008). An inference method for Λ-coalescents is presented in Birkner and Blath (2008).

A population evolving according to the Λ-Fleming-Viot process forms a single mating unit, but real populations are spatially structured. For example they may be subdivided into discrete locations or distributed over a one or two-dimensional continuum. It is to spatial structure that we now turn our attention.

Chapter 6
Spatial Structure

6.1 Subdivided Populations and the Structured Coalescent

Most models of spatially structured populations have the same basic format. The population is assumed to be *subdivided* into *demes*, which one can think of as 'islands' of population. The demes sit at the vertices of a graph and interaction between the subpopulations in different demes is through migration (or more accurately exchange) of individuals along the edges of the graph. The most elementary example is Wright's island model. This is how he introduced it in (Wright (1943)):

> The simplest model is that in which the total population is assumed to be divided into subgroups, each breeding at random within itself, except for a certain proportion of migrants drawn at random from the whole. Since this situation is likely to be approximated in a group of islands, we shall refer to it as the island model.

This corresponds to taking islands at the vertices of a complete graph. More generally one chooses the graph to caricature the spatial environment in which the population evolves. For example populations evolving in a two-dimensional spatial continuum are often approximated by taking the demes to sit at the vertices of $\mathbb{Z}^2$.

To get a feel for the effect that this will have on the genealogical trees for the population we first take a very simple example. Consider a population that is divided into just two demes with migration between the two. This simple model also arises as a model for a single population divided into two genetic types which are in approximate equilibrium in the population, but in which there is mutation between types. The Wright–Fisher model is adapted to this setting as follows:

Definition 6.1 (Wright–Fisher model with migration). A population of size N is structured into two demes, 1 and 2 with population sizes $N_1 = N\omega_1$ in deme 1 and $N_2 = N\omega_2$ in deme 2. Each subpopulation reproduces (independently) according to the neutral Wright–Fisher model except that now, after each reproduction step, a proportion of the population in each deme is exchanged. In other words $\mu_1 N_1$ individuals migrate from deme one to deme two and $\mu_2 N_2$ go the other way. In order to maintain constant population size in each deme, we take $\mu_1 N_1 = \mu_2 N_2$.

We can establish the genealogy of a sample from such a population exactly as in Sect. 5.5. Here things are easier because the population size in each deme is

A. Etheridge, *Some Mathematical Models from Population Genetics*, Lecture Notes in Mathematics 2012, DOI 10.1007/978-3-642-16632-7_6,
© Springer-Verlag Berlin Heidelberg 2011

constant. Because the individuals in a given deme are indistinguishable from one another, the probability that an individual in deme 1 had a parent in deme 2 is just the proportion of individuals in deme 1 after the migration step that had parents in deme 2, namely $\mu_2 N_2 / N_1 = \mu_2 \omega_2 / \omega_1$. Similarly, the probability that an individual in deme 2 had a parent in deme 1 is $\mu_1 \omega_1 / \omega_2$. To obtain a diffusion limit we suppose that $\mu_i = v_i / N$ where $N = N_1 + N_2$ is the total population size and we measure time in units of size N. Since the chance of a migration event and a coalescence event both affecting our ancestral lineages in a single generation is $\mathcal{O}(1/N^2)$, in the diffusion timescale we only see coalescences between lineages in the same deme. Our time unit is the *total population* size, as opposed to the population size in one of the demes, so each pair of lineages currently in deme i, coalesces at instantaneous rate $1/\omega_i$. We are implicitly assuming that $N\omega_i$ is *large* so that we never see multiple mergers. The genealogical trees for this model can then be described by a structured (Kingman) coalescent. As we trace backwards in time

- Ancestral lineages *migrate* from deme one to deme two at rate $v_2 \omega_2 / \omega_1$ and from deme two to deme one at rate $v_1 \omega_1 / \omega_2$.
- Any pair of lineages currently in deme i *coalesces* at instantaneous rate $1/\omega_i$.

Remark 6.2. Notice that the rate of migration of ancestral lineages is weighted by the ratio of the population size in the two demes, just as in Sect. 5.5, so that backwards in time the migration mechanism is biased towards the more populous deme, and, again as in Sect. 5.5, the rate of coalescence within a deme depends on population size there. The analogous result will hold for more general structured populations.

Here we have fixed the total population size in each deme so that different ancestral lineages evolve independently. If we allowed the population size in each deme to fluctuate randomly, then this would no longer be the case. Loosely, knowing that one lineage jumps to a deme tells us that the population size there is probably larger and so other lineages are more likely to jump there too.

Just as we passed to a diffusion approximation from the Wright–Fisher model for a panmictic population, we can also pass to a diffusion approximation for the structured Wright–Fisher model. We assume that the population size in each deme is large enough that the Wright–Fisher diffusion provides a good approximation for the effect of the random resampling due to reproduction. This leads to Kimura's stepping stone model (Kimura (1953)).

Definition 6.3 (Kimura's stepping stone model). We suppose that a population that is distributed across a collection of demes indexed by some set I is also subdivided into two allelic types labelled a and A. The proportion of a-alleles in deme i at time t is denoted by $p_i(t)$. Under Kimura's *stepping stone model*:

$$dp_i = \sum_j m_{ji}(p_j - p_i)dt + \sqrt{\frac{1}{N_e} p_i(1 - p_i)} \, dW_i. \tag{6.1}$$

Here m_{ij} reflects migration between demes and satisfies

$$\sum_{j\neq i} m_{ij} = \sum_{j\neq i} m_{ji} \tag{6.2}$$

(in order to maintain constant population size in each deme). The parameter N_e is the (effective) population size in each deme and the $\{W_i\}_{i\in I}$ are independent standard Brownian motions.

In other words we have a system of interacting Wright–Fisher diffusions. To understand the first term on the right hand side of (6.1), note that type a individuals arrive in deme i at total rate $N_e \sum_j m_{ji} p_j$ and leave at total rate $N_e \sum_j m_{ij} p_i$ and observe that by (6.2)

$$\sum_j m_{ji} p_j - \sum_j m_{ij} p_i = \sum_j m_{ji} p_j - \sum_j m_{ji} p_i = \sum_j m_{ji}(p_j - p_i).$$

Remark 6.4. We can more generally take $N_e(i)$ for the effective population size in deme i, reflecting different population sizes in different demes, but then since we are assuming that the population size in each deme is maintained we must assume that

$$N_e(i) \sum_{j\neq i} m_{ij} = \sum_{j\neq i} N_e(j) m_{ji},$$

and the first term in (6.1) becomes

$$\sum_j \frac{N_e(j)}{N_e(i)} m_{ji}(p_j - p_i).$$

Lemma 6.5. *For a population evolving according to (6.1), the genealogical trees relating a finite sample consisting of n_i individuals from deme i for each $i \in I$ are traced out by the system of coalescing random walks whose evolution is described as follows:*

- *For each $i \in I$, $n_i \mapsto n_i - 1$ at instantaneous rate $\frac{1}{N_e}\binom{n_i}{2}$.*
- *For each $i, j \in I$ with $i \neq j$, $\begin{cases} n_i \mapsto n_i - 1 \\ n_j \mapsto n_j + 1 \end{cases}$ at instantaneous rate $n_i m_{ji}$.*

6.2 Duality

In this section we outline another connection between the stepping stone model and the *structured coalescent* of Lemma 6.5. This is through a powerful technique called the *method of duality*. To illustrate the strengths (and limitations) of the approach, we are going to extend the stepping stone model slightly to incorporate selection.

Definition 6.6 (Kimura's stepping stone model with selection). We suppose that a population that is distributed across a collection of demes indexed by some set I is also subdivided into two allelic types labelled a and A. The proportion of a-alleles in deme i at time t is denoted by $p_i(t)$. Under Kimura's *stepping stone model with selection*

$$dp_i = \sum_j m_{ji}(p_j - p_i)dt + \alpha p_i(1 - p_i)dt + \sqrt{\frac{1}{N_e} p_i(1 - p_i)} dW_i. \qquad (6.3)$$

Here again m_{ij} reflects migration between demes and satisfies

$$\sum_{j \neq i} m_{ij} = \sum_{j \neq i} m_{ji}$$

(in order to maintain constant population size in each deme). The N_e is the (effective) population size in each deme and the $\{W_i\}_{i \in I}$ are independent standard Brownian motions.

The idea of duality is simple. We should like to express the distribution of the process $\underline{p} = (p_i)_{i \in I}$ that we are actually interested in, in our case allele frequencies in different demes, in terms of another (simpler) random variable, $\underline{n}$, that may take values in a completely different state space. The aim is to find a function f for which the following relationship holds:

$$\frac{d}{du} \mathbb{E}\left[f\left(\underline{p}(u), \underline{n}(t - u) \right) \right] = 0, \quad 0 \le u \le t, \qquad (6.4)$$

so that

$$\mathbb{E}\left[f\left(\underline{p}(t), \underline{n}(0) \right) \right] = \mathbb{E}\left[f\left(\underline{p}(0), \underline{n}(t) \right) \right].$$

If, as the second argument of $f(\underline{p}, \underline{n})$ varies, this provides a wide enough class of functions, then this is enough to characterise the distribution of $\underline{p}$. In particular, *existence* of a dual process is often used to prove *uniqueness* (in distribution) of the original process. A good reference is Ethier and Kurtz (1986), see also Etheridge (2000).

It is usually far from evident how to identify a suitable function f, but many models that arise in genetics have *moment duals*. These provide expressions for the moments and mixed moments of the process,

$$\mathbb{E}\left[\prod_{i \in I} p_i^{n_i} \right],$$

where $\underline{n} = (n_i)_{i \in I}$ is a vector with non-negative integer entries, a finite number of which are non-zero. In our dual process we are going to think of n_i as representing a number of 'particles' in deme i. The function f is defined by

$$f(\underline{p}, \underline{n}) = \underline{p}^{\underline{n}} \equiv \prod_{i \in I} p_i^{n_i}$$

and our aim is to find dynamics for the process $\underline{n}(t)$ that guarantee that (6.4) is satisfied. The first step is to calculate $d\underline{p}^{\underline{n}}$ with $\underline{n}$ held fixed. Writing $\underline{e}_i$ for the vector consisting entirely of 0s except for a 1 in the ith position,

$$d\left(\underline{p}^{\underline{n}}\right) = \sum_i n_i \underline{p}^{\underline{n}-\underline{e}_i}\left[\sum_j m_{ji}\left(p_j - p_i\right) + \alpha p_i\left(1 - p_i\right)\right]dt$$
$$+ \sum_i \frac{1}{2N_e}n_i\left(n_i - 1\right)\underline{p}^{\underline{n}-2\underline{e}_i}p_i\left(1 - p_i\right)dt + \sum_i \left(\ldots\right)dW_i.$$

Notice that, because we take the expectation in (6.4), we don't care about the exact form of the martingale term. Rearranging,

$$d\left(\underline{p}^{\underline{n}}\right) = \sum_i n_i \sum_j m_{ji}\left(\underline{p}^{\underline{n}+\underline{e}_j-\underline{e}_i} - \underline{p}^{\underline{n}}\right)dt + \sum_i n_i\alpha\left(\underline{p}^{\underline{n}} - \underline{p}^{\underline{n}+\underline{e}_i}\right)dt$$
$$+ \sum_i \frac{1}{2N_e}n_i\left(n_i - 1\right)\left(\underline{p}^{\underline{n}-\underline{e}_i} - \underline{p}^{\underline{n}}\right)dt + \sum_i \left(\ldots\right)dW_i. \tag{6.5}$$

Our task is to identify dynamics for $\underline{n}(t)$ that ensure that (6.4) holds. To do this, we now think of evaluating $d\underline{p}^{\underline{n}}$ with p held fixed. Notice that since we evaluate $\underline{n}$ at time $t - u$ in (6.4) we pick up an extra minus sign. To cancel the first term in (6.5), particles should migrate according to the time reversal of the random walk that governed the forwards in time evolution of the individuals in our biological population. To cancel the second term we assume that $\alpha \leq 0$. Note that there is no loss of generality in doing so because if we consider $1 - p$ in place of p, that is we look at the proportion of A alleles instead of a alleles, the only effect on (6.6) is to switch the sign of α. If $\alpha < 0$, then the second term will be cancelled by assuming that particles in the dual give birth (split in two) at rate $-\alpha$. Finally, to deal with the last term, we suppose that at instantaneous rate $1/N_e$ each pair of particles currently in deme i coalesces to form a single particle.

We have recovered a spatial version of the ancestral selection graph.

Lemma 6.7. *Suppose that $\underline{p}(t)$ evolves according to the Kimura stepping stone model with selection of Definition 6.6 with $\alpha < 0$ and that the process $\underline{n}$, taking values in $\mathbb{Z}_+^I$ (that is vectors indexed by I with non-negative integer components) and with $\underline{n}(0)$ having only finitely many non-zero components, evolves as follows:*

- $n_i \mapsto n_i + 1$ *at rate* $-\alpha n_i$
- $\begin{cases} n_i \mapsto n_i - 1 \\ n_j \mapsto n_j + 1 \end{cases}$ *at rate* $n_i m_{ji}$
- $n_i \mapsto n_i - 1$ *at rate* $\frac{1}{2N_e}n_i\left(n_i - 1\right)$.

Then we have the duality relationship

$$\mathbb{E}\left[\underline{p}(t)^{\underline{n}(0)}\right] = \mathbb{E}\left[\underline{p}(0)^{\underline{n}(t)}\right].$$

It is easy to explain this result probabilistically. Calculating $\mathbb{E}[\underline{p}(t)^{\underline{n}(0)}]$ is equivalent to asking what is the probability that in a sample consisting of $n_i(0)$ individuals from deme i for each $i \in I$, all individuals are of type a. Just as in the ancestral selection graph of Definition 5.12, the process $\underline{n}(t)$ traces all 'potential' ancestors. The migration and coalescence is what we expect from tracing ancestral lineages of individuals in the sample. The branching of course reflects selection. It is most easily understood in terms of the Moran model with selection of Definition 5.9. The extra 'potential' selective events in the Moran model take place at rate $|\alpha|$. Here (in contrast to Definition 5.9) we are assuming that A has a selective advantage and so if we are to emerge with a type a individual from such a selective event, it must be that both individuals sampled at the event were type a. This happens with probability p^2, hence the branch in the structured coalescent dual – we must check the ancestry of *both* potential parents at such an event.

Remark 6.8. Although the process $\{\underline{n}(t)\}_{t \geq 0}$ has an interpretation in terms of the genealogy of a sample from the population, it is important to remember that the duality relation (6.4) is not enough to guarantee this, c.f. Remark 3.7.

Let's use this duality to try to make some qualitative statements about the long-time behaviour of a population evolving in a two-dimensional habitat. We take $I = \mathbb{Z}^2$ and suppose that migration corresponds to the discrete Laplacian (that is $m_{ij} = 1/4$ if i and j are neighbours and zero otherwise). We consider two separate cases.

First suppose $\alpha < 0$ and to avoid special cases suppose that $0 < p_i(0) < 1$ for all $i \in \mathbb{Z}^2$. Let's calculate

$$\mathbb{E}\left[\underline{p}(t)^{\underline{n}(0)}\right] \qquad \text{as } t \to \infty,$$

for a non-trivial $\underline{n}(0)$. In the dual process of branching and coalescing random walks, branches take place all the time, whereas particles only coalesce when they are in the same site, and the random walk is dispersing them across $\mathbb{Z}^2$, so we expect the number of particles to eventually grow without bound. Irrespective of $\underline{n}(0)$ then,

$$\mathbb{E}\left[\underline{p}(t)^{\underline{n}(0)}\right] = \mathbb{E}\left[\underline{p}(0)^{\underline{n}(t)}\right] \to 0 \qquad \text{as } t \to \infty.$$

Asymptotically, all individuals in our sample will be of type A. This of course makes sense biologically because the type A individuals have a selective advantage.

Now suppose that $\alpha = 0$ so that both alleles are selectively neutral. First we calculate $\mathbb{E}[p_i(t)p_j(t)]$ as $t \to \infty$. To do this, we start the dual process from one particle in site i and one in site j at time zero and see what happens as $t \to \infty$. Now there are no branches any more, just migration and coalescence. The distance between the two particles follows a two-dimensional random walk. Eventually they will come together. When that happens, there is some chance that they will coalesce before they move apart. If they don't coalesce, eventually they will come back together and once again they will have some chance of coalescence. And so on. In finite time they *will* coalesce. Then there will just be a single individual exploring $\mathbb{Z}^2$. The same argument applies for any $\underline{n}(0)$ (with finitely many non-zero components).

Eventually, there will just be a single individual exploring $\mathbb{Z}^2$. Thus

$$\mathbb{E}\left[\underline{p}(t)^{\underline{n}(0)}\right] \to \overline{p} \qquad \text{as } t \to \infty,$$

where $\overline{p}$ is a constant determined by the average initial proportion of a alleles in the population at time zero. How can this happen? Well, only if

$$\underline{p}(t) \xrightarrow{fdd} \begin{cases} \underline{1} \text{ with probability } \overline{p} \\ \underline{0} \text{ with probability } 1 - \overline{p} \end{cases} \qquad \text{as } t \to \infty,$$

where $\underline{1}$ is the vector all of whose entries are 1 and $\underline{0}$ is the vector consisting entirely of 0s and the convergence is in the sense of finite dimensional distributions. So even though neither type has a selective advantage, for large times we expect our sample to consist entirely of a or entirely of A alleles. In the non-spatial setting, since the Wright–Fisher diffusion with no selection or mutation is in the natural scale, the probability that the a allele fixes is its initial frequency (see Lemma 3.14). In the spatial setting, which allele we see in our sample is determined by $\overline{p}$.

6.3 Collapse of Structure

Having established the genealogical trees relating individuals in a sample from a subdivided population one can look for the effect of structure on simple summary statistics of the coalescent trees. Perhaps the best known result is the following.

Lemma 6.9. *Suppose that a population evolves according to Wright's island model with D demes and population size N in each deme. Then the mean coalescence time of two ancestral lineages sampled from within the* same *island is equal to that of two lineages sampled from a panmictic population of size DN independent of the rate of migration between islands.*

Remark 6.10. In fact this result can be extended. For a surprisingly wide range of models of subdivided populations, the mean coalescence time of a sample of two lineages from within a single subpopulation will be equal to that of two individuals sampled from a panmictic population of the same total size, irrespective of the detailed pattern of migration. Conditions to guarantee this can be found in Wilkinson-Herbots (2003).

Proof of Lemma 6.9. Let us write T_{11} for the mean time to coalescence of two lineages sampled at random from within the *same* island and T_{12} for the mean time to coalescence of two lineages sampled from *different* islands. Suppose that the rate of migration of each lineage is m. We condition on the first event to hit the two sampled lineages. If they are in the same island then this can be a migration or a coalescence and happens at exponential rate $2m + 1/N$. If they are in different islands then the event is necessarily a migration. It occurs at rate $2m$ and it can leave the lineages in different islands (with probability $(D-2)/(D-1)$ since only one of the $D-1$ possible targets contains the other lineage) or the same island (with probability $1/(D-1)$). This leads to the linear equations

$$T_{11} = \frac{1}{\frac{1}{N}+2m} + \frac{2m}{\frac{1}{N}+2m}T_{12},$$

$$T_{12} = \frac{1}{2m} + \frac{D-2}{D-1}T_{12} + \frac{1}{D-1}T_{11}.$$

Solving these we obtain

$$T_{11} = ND$$

as required.   □

We can also solve for T_{12} to obtain

$$T_{12} = \frac{D-1}{2m} + ND.$$

This quantity, by contrast, does depend on the migration rate, but if $m \to \infty$ then $T_{12} \to ND$ and the mean time to coalescence behaves as for a panmictic population even if we sample from different demes. One can take this further. Bahlo and Griffiths (2001) find an explicit expression for the Laplace transform of the distribution of the time to the most recent common ancestor of a sample of size two and from this show that, as $m \to \infty$, the whole distribution of the time to the MRCA converges to that of a sample of size two from a panmictic population.

It is natural to ask whether this extends to the genealogical tree of a larger sample from the population. The answer, it turns out, is yes. This is part of a much wider phenomenon in which, because migration and coalescence are happening on different timescales, we see a 'collapse' of structure in the structured coalescent. Nordborg and Krone (2002) summarise the situation beautifully. Here we shall just skim the surface. We consider a population that is subdivided into different states. These could be demes as before or, more generally, age classes, genetic types and so on. If 'migration' (which could be through ageing or mutation for example) between some groups of states is happening on a much faster timescale than coalescence, then the structure associated with those groups collapses and each is replaced, through some sort of averaging procedure, by a single 'metastate'. We already saw an effect like this is Sect. 2.3. When the entire structure collapses, we recover the Kingman coalescent with an effective population size, but one can also recover a structured coalescent in the limit. (We shall see something analogous to this in Sect. 6.5.) To illustrate collapse of structure we consider a very simple example.

Example 6.11. Suppose that our population, which evolves in discrete generations, is divided into two demes with sizes $N_1 = a_1N$ and $N_2 = a_2N$. In each generation, ancestral lineages migrate between demes with strictly positive probabilities β_{12} and β_{21} and we write (γ_1, γ_2) for the stationary distribution of the corresponding random walk. In contrast to Definition 6.1, we do not suppose that β_{ij} is $\mathcal{O}(1/N)$.

Coalescence within demes is with probability $1/N_i = 1/(Na_i)$ in deme i in each generation. Measuring time in units of N generations, the genealogy of a sample from the population converges to a Kingman coalescent in which if there are currently k ancestral lineages, a pair chosen at random will coalesce at rate $c\binom{k}{2}$ where

$$c = \sum_{i=1}^{2} \frac{1}{a_i} \gamma_i^2.$$

We verify this result only when starting from two ancestral lineages and refer to Nordborg and Krone (2002) for a more general result. In this case we can record the possible states of the process of ancestral lineages as

$$\{(1,0),(0,1),(2,0),(1,1),(0,2)\}.$$

Ignoring terms of $\mathcal{O}(1/N^2)$, the backwards in time transition matrix of the process of ancestral lineages can be written as

$$\Pi_N = A + \frac{1}{N}B$$

where the matrix A corresponds to migration of ancestral lineages and the matrix B to coalescence within demes. The key result is the following Lemma which can be found in Möhle (1998).

Lemma 6.12. *Let $t, K \geq 0$ be fixed and let $(c_N)_{N \in \mathbb{N}}$ be a sequence of positive real numbers with $\lim_{N \to \infty} c_N = 0$. Further let $A = (a_{ij})$ be a matrix with $\|A\| \equiv \max_i \sum_j |a_{ij}| = 1$ such that $P = \lim_{n \to \infty} A^n$ exists. Then*

$$\lim_{N \to \infty} \sup_{\|B\| \leq K} \|(A + c_N B)^{\lfloor t/c_N \rfloor} - (P + c_N B)^{\lfloor t/c_N \rfloor}\| = 0.$$

If $(B_N)_{N \in \mathbb{N}}$ is a matrix sequence such that $G = \lim_{N \to \infty} PB_N P$ exists, then

$$\lim_{N \to \infty} (A + c_N B_N)^{\lfloor t/c_N \rfloor} = P - I + e^{tG} \qquad \forall t > 0.$$

This generalises the familiar identity $\lim_{N \to \infty} (I + A/N)^N = e^A$. An easy consequence of this is the following useful theorem.

Theorem 6.13 (Möhle 1998). *Let $X_N = \{X_N(r)\}_{r \in \mathbb{N}_0}$ be a sequence of time homogeneous Markov chains on a probability space $(\Omega, \mathcal{F}, \mathbb{P})$ with the same finite state space S and let Π_N denote the transition matrix of X_N. Assume that the following conditions are satisfied.*

1. *$A = \lim_{N \to \infty} \Pi_N$ exists and $\Pi_N \neq A$ for all sufficiently large N.*
2. *$P = \lim_{n \to \infty} A^n$ exists.*
3. *$G = \lim_{N \to \infty} PB_N P$ exists, where $B_N = (\Pi_N - A)/c_N$ and $c_N = \|\Pi_N - A\|$ for all $N \in \mathbb{N}$.*

If the sequence of initial probability measures $\mathbb{P}_{X_N(0)}$ *converges weakly to some probability measure* μ, *then the finite dimensional distributions of the process* $\{X_N([t/c_N])\}_{t\geq 0}$ *converge to those of a continuous time Markov process* $(X_t)_{t\geq 0}$ *with initial distribution* $X_0 \stackrel{d}{=} \mu$, *transition matrix* $\Pi(t) = P - I + e^{tG}$, $t > 0$, *and infinitesimal generator* G.

Remark 6.14. This is a special case of a general class of results in perturbation theory which are discussed, for example, in Ethier and Kurtz (1986), Chap. 1, Sect. 7.

Since P is a projection, that is $P^2 = P$, we have that

$$P - I + e^{tG} = Pe^{tG} = e^{tG}P.$$

(To see this expand e^{tG}, and hence the left hand side, as a series and note from the definition of G that $PG = G = GP$.) This tells us that the limiting process is obtained by first projecting, using P, onto the stationary distribution of the 'fast process' governed by A and then applying the generator G.

In our example,

$$P = \begin{pmatrix} \gamma_1 & \gamma_2 & 0 & 0 & 0 \\ \gamma_1 & \gamma_2 & 0 & 0 & 0 \\ 0 & 0 & \gamma_1^2 & 2\gamma_1\gamma_2 & \gamma_2^2 \\ 0 & 0 & \gamma_1^2 & 2\gamma_1\gamma_2 & \gamma_2^2 \\ 0 & 0 & \gamma_1^2 & 2\gamma_1\gamma_2 & \gamma_2^2 \end{pmatrix},$$

$c_N = \frac{1}{N}$ and

$$B = \begin{pmatrix} 0 & 0 & 0 & 0 & 0 \\ 0 & 0 & 0 & 0 & 0 \\ \frac{1}{a_1} & 0 & -\frac{1}{a_1} & 0 & 0 \\ 0 & 0 & 0 & 0 & 0 \\ 0 & \frac{1}{a_2} & 0 & 0 & -\frac{1}{a_2} \end{pmatrix}.$$

We can then calculate G as

$$PBP = \begin{pmatrix} 0 & 0 & 0 & 0 & 0 \\ 0 & 0 & 0 & 0 & 0 \\ c\gamma_1 & c\gamma_2 & -c\gamma_1^2 & -2c\gamma_1\gamma_2 & -c\gamma_2^2 \\ c\gamma_1 & c\gamma_2 & -c\gamma_1^2 & -2c\gamma_1\gamma_2 & -c\gamma_2^2 \\ c\gamma_1 & c\gamma_2 & -c\gamma_1^2 & -2c\gamma_1\gamma_2 & -c\gamma_2^2 \end{pmatrix},$$

with

$$c = \frac{\gamma_1^2}{a_1} + \frac{\gamma_2^2}{a_2}.$$

We can collapse states according to the number of lineages to see that we have recovered exactly the Kingman coalescent (up to the time change by c). The assignment of demes to lineages is just by independent sampling according to the stationary distribution of the random walk.   □

In this example, with a fixed and finite number of demes, the result is not really surprising. On the time scale of the coalescence, at any given instant the random walks have probability about γ_1^2 of both being in deme 1 in which case they have instantaneous coalescence rate $1/a_1$ and they have probability about γ_2^2 of both being in deme 2 in which case they coalesce at instantaneous rate $1/a_2$. When we look at larger numbers of lineages, convergence to the coalescent hinges on the exchangeability of lineages. Ancestral lineages have 'forgotten' all about their starting point by the time we see a coalescence event, and so it is equally likely to be any pair of lineages that coalesce. For a general spatial model, we cannot expect this exchangeability for the ancestral lineages of an arbitrary sample. Lineages sampled close together are more likely to coalesce first. However, if coalescence times are long enough, then the lineages have time to 'mix'. Zähle et al. (2005) consider a stepping stone model on a large two-dimensional torus in $\mathbb{Z}^2$. They show that if individuals are sampled uniformly from the torus, then as the side of the torus tends to infinity the genealogy does indeed converge to a Kingman coalescent (with an appropriate effective population size). We shall describe a close analogue of their result in Sect. 6.5.

Collapse of structure can also be seen in island models with large numbers of demes. This is demonstrated in a series of papers by Wakeley and coworkers (e.g., Wakeley (2001), Wakeley and Aliacar (2001)). In contrast to the setting of Nordborg and Krone (2002), the population size, N, in each deme is assumed to be fixed and finite, but the number, D, of demes grows without bound. While within the same deme each pair of lineages coalesces at rate $1/N$, but migration between demes at a rate of $\mathscr{O}(1)$ sends each lineage to a new deme chosen uniformly at random from the $D-1$ available. If the sample size is much smaller than the number of demes, then the chance of landing on a deme that is already occupied by another ancestral lineage is $\mathscr{O}(1/D)$. For large D the history of a sample can then be divided into two phases. During the first *scattering phase*, which is $\mathscr{O}(1)$ generations long, lineages within the same deme will experience a mixture of coalescence and migration to unoccupied demes, until there is at most one lineage in each deme. Never again during the history of the sample will we see more than two lineages in a single deme. During the second *collecting phase*, which is $\mathscr{O}(D)$ generations long, lineages migrate between demes with the possibility of coalescence only when they are in the same deme. Measuring time in units of D generations, we have a tractable ancestral process in which the scattering phase is instantaneous (corresponding to the projection P of Remark 6.14) and the collecting phase is a Kingman coalescent.

6.4 Evolution in a Spatial Continuum and the Pain in the Torus

So far we have concentrated on subdivided populations, but, in reality, many biological populations evolve in a spatial continuum. Wright (1943) and Malécot (1948) considered populations evolving in $\mathbb{R}^1$ and $\mathbb{R}^2$. They make similar assumptions.

Malécot, for example, assumes that (I) individuals are distributed randomly with constant expected density everywhere in space; (II) each individual, independently, produces a Poisson number of offspring with mean one; and (III) each offspring migrates independently, with the displacements being drawn from some distribution $m(x)$, for example a normal distribution. However, as Felsenstein (1975) observed, these assumptions are inconsistent. A population evolving according to (II) and (III) violates (I). In fact, if it is distributed over all of $\mathbb{R}^1$ or $\mathbb{R}^2$ it develops larger and larger clumps separated by greater and greater distances. This is not overcome by working on a torus as then the population dies out. Counteracting this, for example by conditioning the total population size to be a constant N does not overcome the problem of clumping. Felsenstein dubs these problems *'the pain in the torus'*.

Backwards in time, both Wright and Malécot assume that the probability that two individuals have a common parent in the previous generation is a function of their separation (determined by convolving two copies of the distribution $m(x)$) and that if they did not have a common parent their parents' positions are determined by independent copies of $m(x)$. Evidently this (backwards in time) description of the genealogy is not consistent with their forwards in time model for the evolution of the population.[1] So can we find consistent forwards and backwards in time models? In view of the success of the stepping stone model it is natural to use that as a starting point and to try to replace the system of interacting stochastic (ordinary) differential equations by a single stochastic *partial* differential equation. In one spatial dimension this can be achieved by applying the *diffusive rescaling* to the stepping stone model (so that the random walk governing migration of individuals converges to Brownian motion). This results in the limiting equation

$$dp = \frac{1}{2}\Delta p\, dt + \sqrt{\gamma p(1-p)}\, dW, \tag{6.6}$$

where W is now a space-time white noise. This was proved by Shiga (1988), who also established convergence of the system of coalescing random walks that describe the genealogy in the stepping stone model to a system of Brownian motions that coalesce at a rate determined by the local time that they spend together. This generalises work of Nagylaki (1978; 1978) who derived, under the same rescaling, an equation for the correlations between allele frequencies at different locations. In two dimensions Nagylaki showed that the rescaling fails. The equations for the correlations 'blow up' on scales comparable with the distance moved by a single gene over a single generation. Correspondingly, (6.6) has no solution; the white noise is 'too rough'. (See Walsh (1986) for an introduction to stochastic partial differential equations.) Moreover, if one applies the diffusive rescaling to the stepping stone model then one recovers a deterministic heat equation. It is easy to see why by thinking about the genealogical process of coalescing random walks.

---

[1] Wright and Malécot thought about probability of identity in allelic state under an infinitely many alleles mutation model rather than genealogies, but as we saw in Sect. 2.4 the two notions are closely related.

Under the diffusive rescaling the random walks converge to Brownian motions, but two independent Brownian motions evolving in $\mathbb{R}^2$ never meet and so we lose the coalescence. The coalescence is what reflects the noise term (which in turn models the randomness of reproduction) and so with no coalescence we cannot expect any noise.

Remark 6.15. If one modifies (6.6) by replacing the white noise W by a suitable 'coloured' noise, obtained for example by convolving W with a function from $L^2(\mathbb{R}^d)$, then the new equation *does* have a solution. Although at first sight this equation looks natural, it is not, as one might hope, what one obtains by taking a rescaling limit of an individual based model in which parents are chosen from a neighbourhood (rather than the same location). Indeed it is not clear how to obtain it as the limit of *any* individual based model.

Alternatively, instead of modifying the forwards in time stepping stone model, one can try to modify the corresponding (backwards in time) structured coalescent. An obvious approach is to assume that the genealogical trees can be constructed from Brownian motions which coalesce at an instantaneous rate given by a function of their separation. The position of the common ancestor is generally taken to be the midpoint between the two lineages immediately before the coalescence event (although other distributions are of course possible). However, this process of coalescence violates the *consistency* of Remark 2.4. To see this, take the tree corresponding to a sample of size k and consider a subtree of size two. Whenever one of these two ancestral lineages is involved in a coalescence event in the full tree it will jump. We would not see this jump if we modelled the tree relating just two individuals directly. Furthermore, there is no corresponding *forwards* in time model for the evolution of the population.

Wright and Malécot assume an infinitely many alleles mutation model in which, in each generation, each offspring (independently) has a new allelic type with some fixed probability. They find an expression for the probability that two individuals, sampled at distance x apart, have the same allelic type. Although based on inconsistent assumptions, the formula provides an astonishingly good fit to the two-dimensional stepping stone model. This can be seen for example in Fig. 1 of Barton et al. (2002). That paper shows that under certain conditions the Wright–Malécot formula can be extended to continuum population models which incorporate local structure. Over all but very small scales, the resulting probability of identity can be written as a function of three parameters: the *effective dispersal rate*, the *neighbourhood size* and the *local scale*. The difficulty is that there is a shortage of explicit models for which the assumptions underlying this result can be verified and the effective parameters established. Moreover, the formula only applies to samples of size two.

Neighbourhood size is the product of the effective dispersal rate (that is the variance of the Gaussian distribution from which an individual's parent is drawn) and the local population density and gives some measure of how many individuals 'interact' in a given generation. Although the Wright–Malécot formula could in principle be extended to larger samples of well-separated genes, if neighbourhood size is small, multiple coalescences of ancestral lineages could become significant. This observation turned out to be key in writing down a new model which addresses some of the problems identified above.

6.5 The Spatial Λ-Fleming–Viot Process

Recently, in joint work with Nick Barton, we proposed a new framework for modelling populations evolving in a spatial continuum and this will be our final topic. Not only does the proposed framework address some of the issues raised above, including allowing for small neighbourhood size, but it also allows us to explicitly incorporate large-scale extinction-colonisation events into the dynamics of the population. The motivation for this is the basic observation that we made at the beginning of Chap. 5:

> Genetic diversity is orders of magnitude lower than expected from census population size and genetic drift.

While selection certainly plays a rôle in reducing genetic diversity, it is plausible that most of the reduction that we observe relative to the 'null' model of neutral evolution and Kingman's coalescent is due to large scale fluctuations in which the movement and reproductive success of many individuals are correlated. For example climate change has caused extreme extinction and recolonisation events that dominate the demographic history of humans and other species. The new framework provides mathematical models through which to assess the importance of such events relative to some of the other forces that shape genetic diversity.

For simplicity we describe only a particular instance of our approach which can be thought of as a spatial Λ-Fleming–Viot process with genealogical trees determined by a corresponding spatial Λ-coalescent. In this setting, after an extinction event a region is recolonised by the descendants of a single individual. In many settings it would be natural to take a Poisson number of colonists, say, and then the corresponding coalescent model would be a spatial Ξ-coalescent.

The starting point is an individual based model. The resolution of Felsenstein's 'pain in the torus' is that reproduction events (including the large-scale extinction-recolonisation events) are based on a Poisson point process in space. The rate at which a given region of space is affected by such an event does not grow with local population density and this prevents clumping.

Definition 6.16 (Individual based model). We suppose that the population is initially distributed as a Poisson point process in $\mathbb{R}^d$ (with $d = 2$ being the most interesting case). Let λ be a fixed positive constant, $\mu(dr)$ be a measure on $(0, \infty)$ and, for each $r > 0$, let $\nu_r(du)$ be a probability measure on $[0, 1]$ such that

$$\int_0^\infty \int_0^1 u r^d (1 + r^d) \nu_r(du) \mu(dr) < \infty. \tag{6.7}$$

Write $\xi(dr, du) = \mu(dr)v_r(du)$. The dynamics of the population are as follows:

1. Let Π be a Poisson Point Process on $\mathbb{R}_+ \times \mathbb{R}^d \times \mathbb{R}_+ \times (0, 1]$ with rate $dt \otimes dx \otimes \xi(dr, du)$.
2. If (t, x, r, u) is a point of Π, then at time t throw down a ball $B_r(x)$ of radius r and centre x in $\mathbb{R}^d$.
3. If the ball is empty do nothing. If not:

   a. Choose an individual at random from those in $B_r(x)$;
   b. for each individual in $B_r(x)$, independently flip a coin which shows heads with probability u and kill all those individuals with a head;
   c. throw down individuals with the same type as the selected individual (who may now be dead) according to an independent Poisson Point Process with intensity $u\lambda \mathbf{1}_{B_r(x)} dx$.

Regions of space can, and do, become empty in this model, but, because the neighbourhoods affected by different events overlap, an empty region can subsequently be recolonised. Berestycki et al. (2009) show that there is a critical value of λ above which the process survives and below which it dies out. They also check that under condition (6.7) the process described in Definition 6.16 actually exists.

The difficulty with this model is that it is not easy to write down explicitly the genealogical trees relating individuals in a sample from the population. An ancestor is necessarily in a non-empty patch of space and knowing that a region is non-empty gives information about the rate at which it is hit by reproduction events as one traces back in time, but it is hard to find explicit expressions for this effect. We overcome this difficulty by letting $\lambda \to \infty$ so that there are no empty regions of space. At first sight it looks as though we are thereby losing the possibility of small neighbourhood size, but in fact this is not so: by retaining the same reproduction mechanism, in which each individual hit by a reproduction event has probability u of being killed, we retain the signature of finite neighbourhood size. In particular, we can still see multiple coalescences of ancestral lineages.

Remark 6.17. An alternative model of this type, considered in Barton et al. (2010), has a slightly modified reproduction mechanism. It is again based on a spatial Poisson process, but now if an event is centred on the point x, then an individual at y is killed with probability $u(x, y)$, where $u(x, y)$ is a Gaussian kernel centred on x say. A parent is selected by taking a weighted sample from the population immediately before the event, in which individuals are weighted according to their distance from x according to a (possibly different) Gaussian distribution. Offspring, of the same type as the parent, are distributed according to a Poisson point process with intensity $\lambda u(x, y)$. The resulting population model has a Poisson distribution with intensity λ as its stationary distribution.

Let us now describe the limiting model a little more precisely. We suppose that each individual in our population has a *type* taken from a set K (for example $K = [0, 1]$) and a *location* in a space E. For illustration, here we continue to take $E = \mathbb{R}^d$. To each point $x \in E$ and at each time t, the limiting process assigns a probability measure, $\rho(t,x)$ on K. The idea is that the type of an individual sampled from the point x at time t has distribution $\rho(t,x)$. The reproduction mechanism mirrors that for our discrete time model.

Definition 6.18 (Spatial Λ-Fleming–Viot process). The *spatial Λ-Fleming–Viot process*, denoted $\{\rho(t,x,\cdot), x \in \mathbb{R}^d, t \geq 0\}$, specifies a probability measure on the type space K for every $t \geq 0$ and every $x \in \mathbb{R}^d$. With the notation of Definition 6.16, the dynamics of the process are as follows. At every point (t,x,r,u) of the Poisson point process Π we select a point z at random from $B_r(x)$ and a type k at random according to $\rho(t-,z,\cdot)$. For all $y \in B_r(x)$,

$$\rho(t,y,\cdot) = (1-u)\rho(t-,y,\cdot) + u\delta_k.$$

Of course we must impose restrictions on the intensity measure if our process is to exist. To see what these should be, consider first the evolution of the probability measure $\rho(t,x,\cdot)$ defining the distribution of types at the point x. This measure experiences a jump affecting a proportion $y \in A \subseteq [0,1]$ of individuals at x at rate

$$\int_{(0,\infty)} \int_A C_d r^d \nu_r(du)\mu(dr),$$

where C_d is the volume of the unit ball in $\mathbb{R}^d$. By analogy with the Λ-Fleming–Viot process, we should like

$$\tilde{\Lambda}(du) = \int_{(0,\infty)} u^2 r^d \nu_r(du)\mu(dr) \tag{6.8}$$

to define a finite measure on $[0,1]$. In fact, we require a bit more:

$$\Lambda(du) = \int_{(0,\infty)} u r^d \nu_r(du)\mu(dr) \tag{6.9}$$

must define a finite measure on $[0,1]$.

Remark 6.19. Recall from Remark 5.24 that the existence of Λ-coalescents for which the analogue of (6.8) is satisfied, but not that of (6.9), relies on some cancellation of positive and negative jumps. Our need for the stronger condition in the spatial setting reflects the fact that the existence of overlapping neighbourhoods destroys that cancellation.

Of course it is not enough to consider a single point. It has to be possible to 'fit together' the type distributions at different sites in a consistent way and the simplest way to guarantee that we can do this is to ensure the existence of a nice dual process describing, for each $n \in \mathbb{N}$, the distribution of lineages ancestral to a sample of size n from the population. Suppose then that a population evolves according to this model and consider the (backwards in time) dynamics of a *single* ancestral lineage. It evolves in a series of jumps with intensity

$$dt \otimes \int_{(|x|/2,\infty)} \int_{[0,1]} \frac{L_r(x)}{C_d r^d} u\, v_r(du)\mu(dr)dx$$

on $\mathbb{R}_+ \times \mathbb{R}^d$ where $L_r(x)$ is the volume of $B_r(0) \cap B_r(x)$. If we want this to give a well-defined Lévy process, then we require

$$\int_{\mathbb{R}^d}(1 \wedge |x|^2)\left(\int_{(|x|/2,\infty)} \int_{[0,1]} \frac{L_r(x)}{C_d r^d} u\, v_r(du)\mu(dr)\right) dx < \infty. \qquad (6.10)$$

Consider now lineages currently at separation $y \in \mathbb{R}^d$. They will coalesce if they are *both* involved in a replacement event which happens at instantaneous rate

$$\int_{(|y|/2,\infty)} L_r(y)\left(\int_{[0,1]} u^2 v_r(du)\right)\mu(dr). \qquad (6.11)$$

Under condition (6.9), the expressions in (6.10) and (6.11) are both automatically finite. Of course if two ancestral lineages do coalesce, then their common parent is located at a point selected at random from the ball involved in the reproduction event. Conceptually, this is readily extended to multiple lineages (where we will see multiple mergers). Notice that conditional on not having coalesced, the locations of ancestral lineages are *not* independent of one another. This is entirely analogous to the dependence between ancestral lineages in the coalescent for a continuous (finite) linear population suggested by Wilkins and Wakeley (2002) (see Wilkins (2004) for a two-dimensional analogue).

Remark 6.20 (Spatial Λ-coalescent). Evidently the dual process of ancestral lineages is a spatial version of the Λ-coalescent. However, we emphasise that it differs from that studied by Limic and Sturm (2006).

Recall from Sect. 6.3 that the work of Zähle et al. (2005) shows that it makes sense to define a coalescent effective population size (see Remark 2.8) for a uniform sample from a population evolving according to the stepping stone model on a large torus in $\mathbb{Z}^2$. It is natural to ask whether an analogous result holds here and, if so, what the effect of large scale extinction-recolonisation events is on that effective population size. This question is addressed by Barton et al. (2010) and we finish with a description of their result.

Let $\mathbb{T}(L)$ denote the torus of side L in $\mathbb{R}^2$. Suppose that there are two types of event:

1. *Small* events affecting bounded regions;
2. *large* events affecting regions of diameter $\mathcal{O}(L^\alpha)$, for some $0 \le \alpha \le 1$.

The idea is that small events reflect 'ordinary' reproduction, whereas large events model large-scale extinction-recolonisation events. We assume that each ancestral lineage is hit by a small event at rate $\mathcal{O}(1)$, but by a large event at rate $\mathcal{O}(1/\rho(L))$ where $\rho(L) \to \infty$ as $L \to \infty$. We then sample uniformly at random from the whole of $\mathbb{T}(L)$ and ask what happens to the genealogy as $L \to \infty$?

A precise statement can be found in the paper, but here is an outline of the result.

Theorem 6.21 (Barton et al. 2010).

1. *Suppose that $\alpha < 1$. On a suitable timescale the genealogy converges to a Kingman coalescent (with an effective parameter). Depending on $\rho(L)$, the effective population size that determines the timescale can depend on both large and small scale events.*
2. *Suppose that $\alpha = 1$. There are three cases:*
   a. *$\rho(L) \approx L^2$. On timescale $\rho(L)$, the coalescent converges to a spatial Λ-coalescent in which lineages follow independent Brownian motions in between coalescence events.*
   b. *$\rho(L) \approx L^2 \log L$. On timescale $\rho(L)$, the coalescent converges to a (non-spatial) Λ-coalescent in which multiple mergers are due to large events and there can also be a Kingman component reflecting coalescence due to small events.*
   c. *$\rho(L) \gg L^2 \log L$. On a timescale $L^2 \log L$, the coalescent converges to the Kingman coalescent.*

If there are no large events, then in many ways the model looks like a two-dimensional stepping stone model and so, in view of the results of Zähle et al. (2005), it is no surprise to recover the Kingman coalescent. From a biological perspective, what is interesting is the large effect that even rare extinction-recolonisation events can have on the effective population size.

Since for $\alpha = 1$ large scale events cover a non-negligible fraction of the torus, a mathematically much richer picture emerges. If they happen too frequently, then they can affect multiple lineages while the location of those lineages is still correlated with their starting points. If $\rho(L) \approx L^2 \log L$, the positions of ancestral lineages have homogenised over the torus by the time a large event arrives, but lineages have not necessarily yet all coalesced due to small events. Finally, if large events are too rare, then lineages have all coalesced due to small events before we see a large event and so their effect is lost.

6.6 More General Models

One of the attractions of the approach to modelling outlined above is its flexibility. Although we have presented only the simplest form of the spatial Λ-Fleming–Viot process, it can readily be modified to incorporate more realistic biological assumptions. For example, it would be natural to allow for multiple founders after an extinction-recolonisation event and there is no reason to suppose that they are chosen uniformly from the region affected by the event. Equally, we can incorporate selection, recombination, spatial motion of individuals not linked to reproduction and so on.

From a mathematical perspective, even the simplest model reveals a rich structure. For example, by considering a population subdivided into two types, a and A say, with a sufficiently 'sparse', just as for the voter model, if events affect only balls of bounded radius then other than in one spatial dimension one can recover a cluster of superBrownian motion as a rescaling limit for the distribution of a-alleles. By incorporating selection and rescaling suitably, one can obtain the (deterministic) Fisher-KPP equation as a limiting description of allele frequencies. In one dimension one can also recover the stochastic partial differential equation

$$dp = \frac{1}{2}\Delta p\,dt + sp(1-p)dt + \sqrt{\varepsilon p(1-p)}dW, \qquad (6.12)$$

where W is space-time white noise. This equation is the focus of a great deal of current research, but in higher dimensions, which are more biologically relevant, it has no solution. By basing reproduction on regions instead of individuals, we have a natural alternative to (6.12), which makes sense in arbitrary spatial dimensions, from which (6.12) can be recovered as a limit in one spatial dimension, and which arises in a natural way as a limit of an individual based model.

References

Bahlo, M., Griffiths, R.C.: Coalescence time for two genes from a subdivided population. J. Math. Biol. **43**, 397–410 (2001)

Baird, S.J.E., Barton, N.H., Etheridge, A.M.: The distribution of surviving blocks of ancestral genome. Theor. Popul. Biol. **64**, 451–471 (2003)

Barton, N.H.: The effect of hitch-hiking on neutral genealogies. Gen. Res. **72**, 123–133 (1998)

Barton, N.H., Briggs, D.E.G., Eisen, J.A., Goldstein, D.B., Patel, N.H.: Evolution. Cold spring Harbour Press, New York, (2007)

Barton, N.H., Depaulis, F., Etheridge, A.M.: Neutral evolution in spatially continuous populations. Theor. Popul. Biol. **61**, 31–48 (2002)

Barton, N.H., Etheridge, A.M., Sturm, A.K.: Coalescence in a random background. Ann. Appl. Probab. **14**(2), 754–785 (2004)

Barton, N.H., Etheridge, A.M., Véber, A.: A new model for evolution in a spatial continuum. Electron. J. Probab. **15**, 162–216 (2010)

Barton, N.H., Kelleher, J., Etheridge, A.M.: A new model for extinction and recolonization in two dimensions: quantifying phylogeography. Evolution (2010)

Berestycki, J., Berestycki, N., Schweinsberg, J.: Beta-coalescents and continuous stable random trees. Ann. Probab. **35**(5), 1835–1887 (2007)

Berestycki, N: Recent progress in coaelscent theory. Ensaios Matemáticos **16** (2009)

Berestycki, N., Etheridge, A.M., Hutzenthaler, M.: Survival, extinction and ergodicity in a spatially continuous population model. Markov Process. Relat. Fields **15**(3), 265–288 (2009)

Bertoin, J., Le Gall, J.-F.: Stochastic flows associated to a coalescent process. Prob. Theor. Relat. Fields **126**, 261–288 (2003)

Birkner, M., Blath, J.: Computing likelihoods for coalescents with multiple collisions in the infinitely-many-sites model. J. Math. Biol. **57**, 435–465 (2008)

Birkner, M., Blath, J., Capaldo, M., Etheridge, A.M., Möhle, M., Schweinsberg, J., Wakolbinger, A.: Alpha-stable branching and Beta-coalescents. Elect. J. Probab. **10**, 303–325 (2005)

Buri, P.: Gene frequency in small populations of mutant *Drosophila*. Evolution **10**, 367–402 (1956)

Chang, J.T.: Recent common ancestors of all present day individuals. Adv. Appl. Probab. **31**, 1002–1026 (1999)

Darden, T., Kaplan, N.L., Hudson, R.B.: A numerical method for calculating moments of coalescence times in finite populations with selection. J. Math. Biol. **27**(3), 355–368 (1989)

Darwin, C: On the origin of species by means of natural selection. John Murray, London (1859)

Dawson, D.A.: Measure-valued Markov processes. École d'été de probabilités de Saint Flour, vol. 1541, Springer, Berlin (1993)

Donnelly, P.J., Kurtz, T.G.: A countable representation of the Fleming-Viot measure-valued diffusion. Ann. Probab. **24**, 698–742 (1996)

Donnelly, P.J., Kurtz, T.G.: Particle representations for measure-valued population models. Ann. Probab. **27**, 166–205 (1999)

Durrett, R: Stochastic calculus. A practical introduction, CRC, Boca Raton, (1996)

A. Etheridge, *Some Mathematical Models from Population Genetics*, Lecture Notes in Mathematics 2012, DOI 10.1007/978-3-642-16632-7,

Durrett, R., Schweinsberg, J.: Approximating selective sweeps. Theor. Popul. Biol. **66**(2), 129–138 (2004)

Durrett, R., Schweinsberg, J.: A coalescent model for the effect of advantageous mutations on the genealogy of a population. Stoch. Proc. Appl. **115**(10), 1628–1657 (2005)

Eldon, B., Wakeley, J.: Coalescent processes when the distribution of offspring number among individuals is highly skewed. Genetics **172**, 2621–2633 (2006)

Etheridge, A., Pfaffelhuber, P., Wakolbinger, A.: An approximate sampling formula under genetic hitchhiking. Ann. Appl. Probab. **16**(2), 685–729 (2006)

Etheridge, A.M.: An introduction to superprocesses. University lecture notes, vol. 20. Am. Math. Soc., Providence (2000)

Etheridge, A.M., Griffiths, R.C.: A coalescent dual process in a Moran model with genic selection. Theor. Popul. Biol. **75**, 320–330 (2009)

Ethier, S.N., Kurtz, T.G.: Markov processes: characterization and convergence. Wiley, New York (1986)

Ewens, W.J.: The sampling theory of selectively neutral alleles. Theor. Popul. Biol. **3**, 87–112 (1972)

Ewens, W.J.: The concept of the effective population size. Theor. Popul. Biol. **21**, 373–378 (1982)

Ewens, W.J.: Mathematical population genetics i. theoretical introduction. Springer, New York (2004)

Feller, W.: Diffusion processes in genetics. Proc. Second Berkeley Symp. 227–246 (1951)

Feller, W.: Two singular diffusion problems. Ann. Math. **54**, 173–182 (1951)

Felsenstein, J.: A pain in the torus: some difficulties with the model of isolation by distance. Am. Nat. **109**, 359–368 (1975)

Fisher, R.A.: The correlation between relatives on the supposition of Mendelian inheritance. Proc. Roy. Soc. Edinburgh **52**, 399–433 (1918)

Griffiths, R.C.: Ancestral inference from gene trees. In: Slatkin, M., Veuille, M. (eds.) Modern developments in theoretical population genetics: the legacy of Gustave Malécot. Oxford University Press, Oxford (2002)

Haldane, J.B.S.: A mathematical theory of natural and artificial selection. A series of papers beginning in 1924 (1924)

Haldane, J.B.S.: The causes of evolution. Longman, Green and Co., New York (1932)

Kaj, I., Krone, S.M.: The coalescent process in a population with stochastically varying size. J. Appl. Prob. **40**, 33–48 (2003)

Karlin, S., Taylor, H.M.: A second course in stochastic processes. Academic Press, New York (1981)

Kimura, M.: Stepping stone model of population. Ann. Rep. Nat. Inst. Genet. Jpn. **3**, 62–63 (1953)

Kimura, M.: Evolutionary rate at the molecular level. Nature **217**, 624–626 (1968)

Kimura, M.: The neutral theory of molecular evolution. Cambridge University Press, Cambridge (1983)

Kingman, J.F.C.: Random discrete distributions. J. R. Stat. Soc. B **37**, 1–22 (1975)

Kingman, J.F.C.: The population structure associated with the Ewens sampling formula. Theor. Popul. Biol. **11**, 274–283 (1977)

Kingman, J.F.C.: The coalescent. Stoch. Proc. Appl. **13**, 235–248 (1982)

Knight, F.B.: Essentials of Brownian motion and diffusion. Mathematical surveys, vol. 18. Am. Math. Soc., Providence (1981)

Krone, S.M., Neuhauser, C.: Ancestral processes with selection. Theor. Popul. Biol. **51**, 210–237 (1997)

Limic, V., Sturm, A.: The spatial Lambda-coalescent. Electron. J. Probab. **11**(15), 363–393 (2006)

Lynch, M., Conery, J.S.: The origins of genome complexity. Science **302**, 1401–1404 (2003)

Malécot, G.: Les mathématiques de l'hérédité. Masson et Cie, Paris (1948)

Mendel, G.: Versucher über Pflanzenhybriden. Verhandlungen des naturforschenden Vereines in Brünn **Bd. IV für das Jahr, 1865**, 3–47 (1866)

Möhle, M.: A convergence theorem for Markov chains arising in population genetics and the coalescent with selfing. Adv. Appl. Prob. **30**, 493–512 (1998)

Möhle, M.: Total variation distances and rates of convergence for ancestral coalescent processes in exchangeable population models. Adv. Appl. Prob. **32**, 983–993 (2000)

Möhle, M., Sagitov, S.: A classification of coalescent processes for haploid exchangeable models. Ann. Probab. **29**, 1547–1562 (2001)

Moran, P.A.P.: Random processes in genetics. Proc. Camb. Phil. Soc. **54**, 60–71 (1958)

Nagylaki, T.: A diffusion model for geographically structured populations. J. Math. Biol. **6**, 375–382 (1978)

Nagylaki, T.: Random genetic drift in a cline. Proc. Nat. Acad. Sci. USA **75**, 423–426 (1978)

Neuhauser, C., Krone, S.M.: Genealogies of samples in models with selection. Genetics **145**, 519–534 (1997)

Nordborg, M., Donnelly, P.: The coalescent process with selfing. Genetics **146**, 1185–1195 (1997)

Nordborg, M., Krone, S.M.: Separation of timescales and convergence to the coalescent in structured populations. In: Slatkin, M., Veuille, M. (eds.) Modern developments in theoretical population genetics: the legacy of Gustave Malécot. Oxford University Press, Oxford (2002)

Pitman, J.: Coalescents with multiple collisions. Ann. Probab. **27**, 1870–1902 (1999)

Sagitov, S.: The general coalescent with asynchronous mergers of ancestral lines. J. Appl. Probab. **26**, 1116–1125 (1999)

Sargsyan, O., Wakeley, J.: A coalescent processes with simultaneous mutliple mergers for approximating the gene genealogies of many marine organisms. Theor. Popul. Biol. **74**, 104–114 (2008)

Schweinsberg, J.: Coalescents with simultaneous multiple collisions. Electron. J. Probab. **5**, 1–50 (2000)

Schweinsberg, J.: Coalescents obtained from supercritical Galton-Watson processes. Stoch. Proc. Appl. **106**, 107–139 (2003)

Schweinsberg, J., Durrett, R.: Random partitions approximating the coalescence of lineages during a selective sweep. Ann. Appl. Probab. **15**(3), 1591–1651 (2005)

Shiga, T.: Stepping stone models in population genetics and population dynamics. In: Albeverio, S., et al. (eds.) Stochastic processes in physics and engineering. D Reidel Publishing Company, Dordrecht (1988)

Sjödin, P., Kaj, I., Krone, S., Lascoux, M., Nordborg, M.: On the meaning and existence of an effective population size. Genetics **169**, 1061–1070 (2005)

Stroock, D.W., Varadhan, S.R.S.: Multidimensional diffusion processes. Springer, Berlin (1979)

Taylor, J.E.: The genealogical consequences of fecundity variance polymorphism. Genetics **182**(3), 813–837 (2009)

Wakeley, J.: The coalescent in an island model of population subdivision with variation among demes. Theor. Popul. Biol. **59**, 133–144 (2001)

Wakeley, J., Aliacar, N.: Gene genealogies in a metapopulation. Genetics **159**, 893–905 (2001)

Walsh, J.B.: An introduction to stochastic partial differential equations. École d'été de probabilités de Saint Flour, vol. 1180. Springer, New York (1986)

Watanabe, H., Motoo, M.: Ergodic property of recurrent diffusion processes. J. Math. Soc. Jpn. **10**, 271–286 (1958)

Watterson, G.A.: Reversibility and the age of an allele II. Two-allele models with selection and mutation. Theor. Popul. Biol. **12**(2), 179–196 (1977)

Wilkins, J.F.: A separation of timescales approach to the coalescent in a continuous population. Genetics **168**, 2227–2244 (2004)

Wilkins, J.F., Wakeley, J.: The coalescent in a continuous, finite, linear population. Genetics **161**, 873–888 (2002)

Wilkinson-Herbots, H.M.: Coalescence time and F_{ST} values in subdivided populations with symmetric structure. Adv. Appl. Prob. **35**, 665–690 (2003)

Wright, S.: Isolation by distance. Genetics **28**, 114–138 (1943)

Zähle, I., Cox, J.T., Durrett, R.: The stepping stone model ii: genealogies and the infinite sites model. Ann. Appl. Probab. **15**, 671–699 (2005)

List of Participants

39th Probability Summer School, Saint-Flour, France

July 5–18, 2009

Lecturers

Robert ADLER	Technion Inst. Technology, Haifa, Israel
Alison ETHERIDGE	University of Oxford, UK

Participants

Mamadou BA	Univ. de Provence, Marseille, F
Vincent BANSAYE	Univ. Pierre et Marie Curie, Paris, F
David BASCOMPTE	Univ. Autònoma Barcelona, Spain
Mireia BESALÚ	Univ. Barcelona, Spain
Hermine BIERMÉ	Univ. Paris Descartes, F
Max BRUGGER	Oregon State University, Corvallis, USA
Valentina CAMMAROTA	Univ. Roma "Sapienza", Italy
Alexandra CHRONOPOULOU	Purdue University, West Lafayette, USA
Charles CUTHBERTSON	Morgan Stanley, London, UK
François D'HAUTEFEUILLE	London, UK
Mirko D'OVIDIO	Univ. Roma "Sapienza", Italy
Francisco Javier DELGADO	Univ. Barcelona, Spain
Yogeshwaran DHANDAPANI	École Normale Supérieure, Paris, F
Roland DIEL	Univ. Orléans, F
Leif DÖRING	TU Berlin, Germany
Richard EDEN	Purdue University, West Lafayette, USA
Anne ESTRADE	Univ. Orléans, F
Mikael FALCONNET	Univ. J. Fourier, Grenoble, F
Benjamin FAVETTO	Univ. Paris Descartes, F
Antoine GERBAUD	Univ. J. Fourier, Grenoble, F

Raouf GHOMRASNI   African Inst. Math. Sciences, South Africa
Priscilla GREENWOOD   Arizona State University, Tempe, USA
Simona GRUSEA   Univ. de Provence, Marseille, F
M. O. HAJI MIRSADEGHI   École Normale Supérieure, Paris, F
Olivier HÉNARD   CERMICS, Marne-la-Vallée, F
Marie KRATZ   Univ. Paris Descartes, F
Krzysztof LATUSZYNSKI   Univ. Warwick, UK
Mickaël LAUNAY   Univ. de Provence, Marseille, F
Jean-François LE GALL   Univ. Paris-Sud, Orsay, F
Sophie LEMAIRE   Univ. Paris-Sud, Orsay, F
Philippe MARCHAL   École Normale Supérieure, Paris, F
Takahiro MARUKI   Arizona State University, Tempe, USA
Grégory MIERMONT   École Normale Supérieure, Paris, F
Alina NICOLAIE   Leiden Univ. Medical Center, NL
Bernardo NIPOTI   Univ. Pavia, Italy
Raoul NORMAND   Univ. Pierre et Marie Curie, Paris, F
Piotr NOWAK   Univ. Wroclaw, Poland
Janosch ORTMANN   Univ. Warwick, UK
Todd PARSONS   Univ. Pennsylvania, Philadelphia, USA
Jean PICARD   Univ. Blaise Pascal, Clermont-Ferrand, F
Balakrishna PRABHU   LAAS, Toulouse, F
Habib SAADI   Oxford Univ., UK
Majid SALAMAT   Univ. de Provence, Marseille, F
Laurent SERLET   Univ. Blaise Pascal, Clermont-Ferrand, F
Florian SIMATOS   INRIA Paris-Rocquencourt, F
Kinga SIPOS   Babes-Bolyai Univ., Cluj Napoca, Romania
Arno SIRI-JÉGOUSSE   Univ. Paris Descartes, F
Katharina SUROVCIK   Univ. Freiburg, Germany
Jonathan TAYLOR   Stanford Univ., USA
Marie THÉRET   École Normale Supérieure, Paris, F
Amandine VÉBER   Univ. Paris-Sud, Orsay, F
Juan VIQUEZ   Purdue University, West Lafayette, USA
Guillaume VOISIN   Univ. Orléans, F
Joseph ZADEH   Purdue University, West Lafayette, USA
Lorenzo ZAMBOTTI   Univ. Pierre et Marie Curie, Paris, F

Programme of the School

Main lectures

Robert Adler — Topological complexity of smooth random functions

Alison Etheridge — Some mathematical models from population genetics

Short lectures

Vincent Bansaye — Large deviations for branching processes in random environment

David Bascompte — On the convergence toward two Gaussian processes constructed from a unique Poisson process

Hermine Biermé — Shot noise processes and crossings

Valentina Cammarota — Pseudo-processes: joint distribution of the pseudo-process and its sojourn time in $(0, +\infty)$

Alexandra Chronopoulou — Variations and Hurst index estimation for a Rosenblatt process using longer filters

Charles Cuthberson — Fixation probability for competing selective sweeps

François d'Hautefeuille — Dimer model and Monte Carlo simulation

Yogeshwaran Dhandapani — Directionally convex ordering of random measures, shot-noise fields and some applications to wireless communications

Roland Diel — Local time of a diffusion in Brownian environment

Leif Döring — Longtime behaviour of symbiotic branching processes

Mikael Falconnet — Phylogenetic distances for neighbour dependent substitution processes

Benjamin Favetto — On the asymptotic variance in the central limit theorem for particle filters

Antoine Gerbaud — A random difference equation arising from a new model of stochastic network

Raouf Ghomrasni — Recent results on local time of semimartingales

Mir Omid Haji Mirsadeghi — First passage percolation in SINR random graph

Krzysztof Latuszyński	Simulating events of unknown probabilities via reverse time martingales
Philippe Marchal	Small time asymptotics for Lévy processes
Takahiro Maruki	A stochastic model of the hitchhiking effect
Grégory Miermont	Discrete splitting models and their scaling limits
Alina Nicolaie	Vertical modelling: a pattern mixture approach for competing risks modelling
Raoul Normand	A model for sexed coagulation
Todd Parsons	Some results for density dependent population genetics
Florian Simatos	Spatial homogenization in a stochastic network with mobility
Kinga Sipos	Optimization in cash management
Katharina Surovcik	Profiles for a class of binary random trees
Marie Théret	Maximal flow through a domain of R^d in first passage percolation
Amandine Véber	Branching Brownian motion among random obstacles
Guillaume Voisin	Fragmentation of a Lévy continuum random tree

Index

Lecture Notes in Mathematics

For information about earlier volumes
please contact your bookseller or Springer
LNM Online archive: springerlink.com

Vol. 1867: J. Sneyd (Ed.), Tutorials in Mathematical Biosciences II. Mathematical Modeling of Calcium Dynamics and Signal Transduction. (2005)

Vol. 1868: J. Jorgenson, S. Lang, $Pos_n(R)$ and Eisenstein Series. (2005)

Vol. 1869: A. Dembo, T. Funaki, Lectures on Probability Theory and Statistics. Ecole d'Eté de Probabilités de Saint-Flour XXXIII-2003. Editor: J. Picard (2005)

Vol. 1870: V.I. Gurariy, W. Lusky, Geometry of Mntz Spaces and Related Questions. (2005)

Vol. 1871: P. Constantin, G. Gallavotti, A.V. Kazhikhov, Y. Meyer, S. Ukai, Mathematical Foundation of Turbulent Viscous Flows, Martina Franca, Italy, 2003. Editors: M. Cannone, T. Miyakawa (2006)

Vol. 1872: A. Friedman (Ed.), Tutorials in Mathematical Biosciences III. Cell Cycle, Proliferation, and Cancer (2006)

Vol. 1873: R. Mansuy, M. Yor, Random Times and Enlargements of Filtrations in a Brownian Setting (2006)

Vol. 1874: M. Yor, M. Émery (Eds.), In Memoriam Paul-Andr Meyer - Sminaire de Probabilits XXXIX (2006)

Vol. 1875: J. Pitman, Combinatorial Stochastic Processes. Ecole d'Et de Probabilits de Saint-Flour XXXII-2002. Editor: J. Picard (2006)

Vol. 1876: H. Herrlich, Axiom of Choice (2006)

Vol. 1877: J. Steuding, Value Distributions of L-Functions (2007)

Vol. 1878: R. Cerf, The Wulff Crystal in Ising and Percolation Models, Ecole d'Et de Probabilités de Saint-Flour XXXIV-2004. Editor: Jean Picard (2006)

Vol. 1879: G. Slade, The Lace Expansion and its Applications, Ecole d'Et de Probabilits de Saint-Flour XXXIV-2004. Editor: Jean Picard (2006)

Vol. 1880: S. Attal, A. Joye, C.-A. Pillet, Open Quantum Systems I, The Hamiltonian Approach (2006)

Vol. 1881: S. Attal, A. Joye, C.-A. Pillet, Open Quantum Systems II, The Markovian Approach (2006)

Vol. 1882: S. Attal, A. Joye, C.-A. Pillet, Open Quantum Systems III, Recent Developments (2006)

Vol. 1883: W. Van Assche, F. Marcellàn (Eds.), Orthogonal Polynomials and Special Functions, Computation and Application (2006)

Vol. 1884: N. Hayashi, E.I. Kaikina, P.I. Naumkin, I.A. Shishmarev, Asymptotics for Dissipative Nonlinear Equations (2006)

Vol. 1885: A. Telcs, The Art of Random Walks (2006)

Vol. 1886: S. Takamura, Splitting Deformations of Degenerations of Complex Curves (2006)

Vol. 1887: K. Habermann, L. Habermann, Introduction to Symplectic Dirac Operators (2006)

Vol. 1888: J. van der Hoeven, Transseries and Real Differential Algebra (2006)

Vol. 1889: G. Osipenko, Dynamical Systems, Graphs, and Algorithms (2006)

Vol. 1890: M. Bunge, J. Funk, Singular Coverings of Toposes (2006)

Vol. 1891: J.B. Friedlander, D.R. Heath-Brown, H. Iwaniec, J. Kaczorowski, Analytic Number Theory, Cetraro, Italy, 2002. Editors: A. Perelli, C. Viola (2006)

Vol. 1892: A. Baddeley, I. Bárány, R. Schneider, W. Weil, Stochastic Geometry, Martina Franca, Italy, 2004. Editor: W. Weil (2007)

Vol. 1893: H. Hanßmann, Local and Semi-Local Bifurcations in Hamiltonian Dynamical Systems, Results and Examples (2007)

Vol. 1894: C.W. Groetsch, Stable Approximate Evaluation of Unbounded Operators (2007)

Vol. 1895: L. Molnár, Selected Preserver Problems on Algebraic Structures of Linear Operators and on Function Spaces (2007)

Vol. 1896: P. Massart, Concentration Inequalities and Model Selection, Ecole d'Été de Probabilités de Saint-Flour XXXIII-2003. Editor: J. Picard (2007)

Vol. 1897: R. Doney, Fluctuation Theory for Lévy Processes, Ecole d'Été de Probabilités de Saint-Flour XXXV-2005. Editor: J. Picard (2007)

Vol. 1898: H.R. Beyer, Beyond Partial Differential Equations, On linear and Quasi-Linear Abstract Hyperbolic Evolution Equations (2007)

Vol. 1899: Séminaire de Probabilités XL. Editors: C. Donati-Martin, M. Émery, A. Rouault, C. Stricker (2007)

Vol. 1900: E. Bolthausen, A. Bovier (Eds.), Spin Glasses (2007)

Vol. 1901: O. Wittenberg, Intersections de deux quadriques et pinceaux de courbes de genre 1, Intersections of Two Quadrics and Pencils of Curves of Genus 1 (2007)

Vol. 1902: A. Isaev, Lectures on the Automorphism Groups of Kobayashi-Hyperbolic Manifolds (2007)

Vol. 1903: G. Kresin, V. Maz'ya, Sharp Real-Part Theorems (2007)

Vol. 1904: P. Giesl, Construction of Global Lyapunov Functions Using Radial Basis Functions (2007)

Vol. 1905: C. Prévôt, M. Röckner, A Concise Course on Stochastic Partial Differential Equations (2007)

Vol. 1906: T. Schuster, The Method of Approximate Inverse: Theory and Applications (2007)

Vol. 1907: M. Rasmussen, Attractivity and Bifurcation for Nonautonomous Dynamical Systems (2007)

Vol. 1908: T.J. Lyons, M. Caruana, T. Lévy, Differential Equations Driven by Rough Paths, Ecole d'Été de Probabilités de Saint-Flour XXXIV-2004 (2007)

Vol. 1909: H. Akiyoshi, M. Sakuma, M. Wada, Y. Yamashita, Punctured Torus Groups and 2-Bridge Knot Groups (I) (2007)

Vol. 1910: V.D. Milman, G. Schechtman (Eds.), Geometric Aspects of Functional Analysis. Israel Seminar 2004-2005 (2007)

Vol. 1911: A. Bressan, D. Serre, M. Williams, K. Zumbrun, Hyperbolic Systems of Balance Laws. Cetraro, Italy 2003. Editor: P. Marcati (2007)

Vol. 1912: V. Berinde, Iterative Approximation of Fixed Points (2007)

Vol. 1913: J.E. Marsden, G. Misiołek, J.-P. Ortega, M. Perlmutter, T.S. Ratiu, Hamiltonian Reduction by Stages (2007)

Vol. 1914: G. Kutyniok, Affine Density in Wavelet Analysis (2007)

Vol. 1915: T. Bıyıkoğlu, J. Leydold, P.F. Stadler, Laplacian Eigenvectors of Graphs. Perron-Frobenius and Faber-Krahn Type Theorems (2007)

Vol. 1916: C. Villani, F. Rezakhanlou, Entropy Methods for the Boltzmann Equation. Editors: F. Golse, S. Olla (2008)

Vol. 1917: I. Veselić, Existence and Regularity Properties of the Integrated Density of States of Random Schrdinger (2008)

Vol. 1918: B. Roberts, R. Schmidt, Local Newforms for GSp(4) (2007)

Vol. 1969: B. Roynette, M. Yor, Penalising Brownian Paths (2009)

Vol. 1970: M. Biskup, A. Bovier, F. den Hollander, D. Ioffe, F. Martinelli, K. Netočný, F. Toninelli, Methods of Contemporary Mathematical Statistical Physics. Editor: R. Kotecký (2009)

Vol. 1971: L. Saint-Raymond, Hydrodynamic Limits of the Boltzmann Equation (2009)

Vol. 1972: T. Mochizuki, Donaldson Type Invariants for Algebraic Surfaces (2009)

Vol. 1973: M.A. Berger, L.H. Kauffmann, B. Khesin, H.K. Moffatt, R.L. Ricca, De W. Sumners, Lectures on Topological Fluid Mechanics. Cetraro, Italy 2001. Editor: R.L. Ricca (2009)

Vol. 1974: F. den Hollander, Random Polymers: École d'Été de Probabilités de Saint-Flour XXXVII – 2007 (2009)

Vol. 1975: J.C. Rohde, Cyclic Coverings, Calabi-Yau Manifolds and Complex Multiplication (2009)

Vol. 1976: N. Ginoux, The Dirac Spectrum (2009)

Vol. 1977: M.J. Gursky, E. Lanconelli, A. Malchiodi, G. Tarantello, X.-J. Wang, P.C. Yang, Geometric Analysis and PDEs. Cetraro, Italy 2001. Editors: A. Ambrosetti, S.-Y.A. Chang, A. Malchiodi (2009)

Vol. 1978: M. Qian, J.-S. Xie, S. Zhu, Smooth Ergodic Theory for Endomorphisms (2009)

Vol. 1979: C. Donati-Martin, M. Émery, A. Rouault, C. Stricker (Eds.), Séminaire de Probablitiés XLII (2009)

Vol. 1980: P. Graczyk, A. Stos (Eds.), Potential Analysis of Stable Processes and its Extensions (2009)

Vol. 1981: M. Chlouveraki, Blocks and Families for Cyclotomic Hecke Algebras (2009)

Vol. 1982: N. Privault, Stochastic Analysis in Discrete and Continuous Settings. With Normal Martingales (2009)

Vol. 1983: H. Ammari (Ed.), Mathematical Modeling in Biomedical Imaging I. Electrical and Ultrasound Tomographies, Anomaly Detection, and Brain Imaging (2009)

Vol. 1984: V. Caselles, P. Monasse, Geometric Description of Images as Topographic Maps (2010)

Vol. 1985: T. Linß, Layer-Adapted Meshes for Reaction-Convection-Diffusion Problems (2010)

Vol. 1986: J.-P. Antoine, C. Trapani, Partial Inner Product Spaces. Theory and Applications (2009)

Vol. 1987: J.-P. Brasselet, J. Seade, T. Suwa, Vector Fields on Singular Varieties (2010)

Vol. 1988: M. Broué, Introduction to Complex Reflection Groups and Their Braid Groups (2010)

Vol. 1989: I.M. Bomze, V. Demyanov, Nonlinear Optimization. Cetraro, Italy 2007. Editors: G. di Pillo, F. Schoen (2010)

Vol. 1990: S. Bouc, Biset Functors for Finite Groups (2010)

Vol. 1991: F. Gazzola, H.-C. Grunau, G. Sweers, Polyharmonic Boundary Value Problems (2010)

Vol. 1992: A. Parmeggiani, Spectral Theory of Non-Commutative Harmonic Oscillators: An Introduction (2010)

Vol. 1993: P. Dodos, Banach Spaces and Descriptive Set Theory: Selected Topics (2010)

Vol. 1994: A. Baricz, Generalized Bessel Functions of the First Kind (2010)

Vol. 1995: A.Y. Khapalov, Controllability of Partial Differential Equations Governed by Multiplicative Controls (2010)

Vol. 1996: T. Lorenz, Mutational Analysis. A Joint Framework for Cauchy Problems *In* and *Beyond* Vector Spaces (2010)

Vol. 1997: M. Banagl, Intersection Spaces, Spatial Homology Truncation, and String Theory (2010)

Vol. 1998: M. Abate, E. Bedford, M. Brunella, T.-C. Dinh, D. Schleicher, N. Sibony, Holomorphic Dynamical Systems. Cetraro, Italy 2008. Editors: G. Gentili, J. Guenot, G. Patrizio (2010)

Vol. 1999: H. Schoutens, The Use of Ultraproducts in Commutative Algebra (2010)

Vol. 2000: H. Yserentant, Regularity and Approximability of Electronic Wave Functions (2010)

Vol. 2001: T. Duquesne, O. Reichmann, K.-i. Sato, C. Schwab, Lévy Matters I. Editors: O.E. Barndorff-Nielson, J. Bertoin, J. Jacod, C. Klüppelberg (2010)

Vol. 2002: C. Pötzsche, Geometric Theory of Discrete Nonautonomous Dynamical Systems (2010)

Vol. 2003: A. Cousin, S. Crépey, O. Guéant, D. Hobson, M. Jeanblanc, J.-M. Lasry, J.-P. Laurent, P.-L. Lions, P. Tankov, Paris-Princeton Lectures on Mathematical Finance 2010. Editors: R.A. Carmona, E. Cinlar, I. Ekeland, E. Jouini, J.A. Scheinkman, N. Touzi (2010)

Vol. 2004: K. Diethelm, The Analysis of Fractional Differential Equations (2010)

Vol. 2005: W. Yuan, W. Sickel, D. Yang, Morrey and Campanato Meet Besov, Lizorkin and Triebel (2011)

Vol. 2006: C. Donati-Martin, A. Lejay, W. Rouault (Eds.), Séminaire de Probabilités XLIII (2011)

Vol. 2007: E. Bujalance, F.J. Cirre, J.M. Gamboa, G. Gromadzki, Symmetries of Compact Riemann Surfaces (2010)

Vol. 2008: P.F. Baum, G. Cortiñas, R. Meyer, R. Sánchez-García, M. Schlichting, B. Toën, Topics in Algebraic and Topological K-Theory. Editor: G. Cortiñas (2011)

Vol. 2009: J.-L. Colliot-Thélène, P.S. Dyer, P. Vojta, Arithmetic Geometry. Cetraro, Italy 2007. Editors: P. Corvaja, C. Gasbarri (2011)

Vol. 2010: A. Farina, A. Klar, R.M.M. Mattheij, A. Mikelić, N. Siedow, Mathematical Models in the Manufacturing of Glass. Cetraro, Italy 2008. Editor: A. Fasano (2011)

Vol. 2011: B. Andrews, C. Hopper, The Ricci Flow in Riemannian Geometry (2011)

Vol. 2012: A. Etheridge, Some Mathematical Models from Population Genetics. École d'Été de Probabilités de Saint-Flour XXXIX-2009 (2011)

Recent Reprints and New Editions

Vol. 1702: J. Ma, J. Yong, Forward-Backward Stochastic Differential Equations and their Applications. 1999 – Corr. 3rd printing (2007)

Vol. 830: J.A. Green, Polynomial Representations of GL_n, with an Appendix on Schensted Correspondence and Littelmann Paths by K. Erdmann, J.A. Green and M. Schoker 1980 – 2nd corr. and augmented edition (2007)

Vol. 1693: S. Simons, From Hahn-Banach to Monotonicity (Minimax and Monotonicity 1998) – 2nd exp. edition (2008)

Vol. 470: R.E. Bowen, Equilibrium States and the Ergodic Theory of Anosov Diffeomorphisms. With a preface by D. Ruelle. Edited by J.-R. Chazottes. 1975 – 2nd rev. edition (2008)

Vol. 523: S.A. Albeverio, R.J. Høegh-Krohn, S. Mazzucchi, Mathematical Theory of Feynman Path Integral. 1976 – 2nd corr. and enlarged edition (2008)

Vol. 1764: A. Cannas da Silva, Lectures on Symplectic Geometry 2001 – Corr. 2nd printing (2008)

LECTURE NOTES IN MATHEMATICS   🐎 **Springer**

Edited by J.-M. Morel, F. Takens, B. Teissier, P.K. Maini

Editorial Policy (for the publication of monographs)

1. Lecture Notes aim to report new developments in all areas of mathematics and their applications - quickly, informally and at a high level. Mathematical texts analysing new developments in modelling and numerical simulation are welcome.

   Monograph manuscripts should be reasonably self-contained and rounded off. Thus they may, and often will, present not only results of the author but also related work by other people. They may be based on specialised lecture courses. Furthermore, the manuscripts should provide sufficient motivation, examples and applications. This clearly distinguishes Lecture Notes from journal articles or technical reports which normally are very concise. Articles intended for a journal but too long to be accepted by most journals, usually do not have this "lecture notes" character. For similar reasons it is unusual for doctoral theses to be accepted for the Lecture Notes series, though habilitation theses may be appropriate.

2. Manuscripts should be submitted either online at www.editorialmanager.com/lnm to Springer's mathematics editorial in Heidelberg, or to one of the series editors. In general, manuscripts will be sent out to 2 external referees for evaluation. If a decision cannot yet be reached on the basis of the first 2 reports, further referees may be contacted: The author will be informed of this. A final decision to publish can be made only on the basis of the complete manuscript, however a refereeing process leading to a preliminary decision can be based on a pre-final or incomplete manuscript. The strict minimum amount of material that will be considered should include a detailed outline describing the planned contents of each chapter, a bibliography and several sample chapters.

   Authors should be aware that incomplete or insufficiently close to final manuscripts almost always result in longer refereeing times and nevertheless unclear referees' recommendations, making further refereeing of a final draft necessary.

   Authors should also be aware that parallel submission of their manuscript to another publisher while under consideration for LNM will in general lead to immediate rejection.

3. Manuscripts should in general be submitted in English. Final manuscripts should contain at least 100 pages of mathematical text and should always include

   – a table of contents;
   – an informative introduction, with adequate motivation and perhaps some historical remarks: it should be accessible to a reader not intimately familiar with the topic treated;
   – a subject index: as a rule this is genuinely helpful for the reader.

   For evaluation purposes, manuscripts may be submitted in print or electronic form (print form is still preferred by most referees), in the latter case preferably as pdf- or zipped ps-files. Lecture Notes volumes are, as a rule, printed digitally from the authors' files. To ensure best results, authors are asked to use the LaTeX2e style files available from Springer's web-server at:

   ftp://ftp.springer.de/pub/tex/latex/svmonot1/ (for monographs) and
   ftp://ftp.springer.de/pub/tex/latex/svmultt1/ (for summer schools/tutorials).
   Additional technical instructions, if necessary, are available on request from: lnm@springer.com.

4. Careful preparation of the manuscripts will help keep production time short besides ensuring satisfactory appearance of the finished book in print and online. After acceptance of the manuscript authors will be asked to prepare the final LaTeX source files and also the corresponding dvi-, pdf- or zipped ps-file. The LaTeX source files are essential for producing the full-text online version of the book (see http://www.springerlink.com/openurl.asp?genre=journal&issn=0075-8434 for the existing online volumes of LNM).

   The actual production of a Lecture Notes volume takes approximately 12 weeks.

5. Authors receive a total of 50 free copies of their volume, but no royalties. They are entitled to a discount of 33.3% on the price of Springer books purchased for their personal use, if ordering directly from Springer.

6. Commitment to publish is made by letter of intent rather than by signing a formal contract. Springer-Verlag secures the copyright for each volume. Authors are free to reuse material contained in their LNM volumes in later publications: a brief written (or e-mail) request for formal permission is sufficient.

Addresses:

Professor J.-M. Morel, CMLA,
École Normale Supérieure de Cachan,
61 Avenue du Président Wilson, 94235 Cachan Cedex, France
E-mail: Jean-Michel.Morel@cmla.ens-cachan.fr

Professor F. Takens, Mathematisch Instituut,
Rijksuniversiteit Groningen, Postbus 800,
9700 AV Groningen, The Netherlands
E-mail: F.Takens@rug.nl

Professor B. Teissier, Institut Mathématique de Jussieu,
UMR 7586 du CNRS, Équipe "Géométrie et Dynamique",
175 rue du Chevaleret,
75013 Paris, France
E-mail: teissier@math.jussieu.fr

For the "Mathematical Biosciences Subseries" of LNM:

Professor P.K. Maini, Center for Mathematical Biology,
Mathematical Institute, 24-29 St Giles,
Oxford OX1 3LP, UK
E-mail: maini@maths.ox.ac.uk

Springer, Mathematics Editorial, Tiergartenstr. 17,
69121 Heidelberg, Germany,
Tel.: +49 (6221) 487-259
Fax: +49 (6221) 4876-8259
E-mail: lnm@springer.com